MYSTERIOUS MEDICINE

LITERATURE AND MEDICINE

Michael Blackie, Editor • Carol Donley and Martin Kohn, Founding Editors

1 *Literature and Aging: An Anthology*
Edited by Martin Kohn, Carol
Donley, and Delese Wear

2 *The Tyranny of the Normal: An
Anthology*
Edited by Carol Donley and Sheryl
Buckley

3 *What's Normal? Narratives of
Mental and Emotional Disorders*
Edited by Carol Donley and Sheryl
Buckley

4 *Recognitions: Doctors and Their
Stories*
Edited by Carol Donley and Mar-
tin Kohn

5 *Chekhov's Doctors: A Collection of
Chekhov's Medical Tales*
Edited by Jack Coulehan

6 *Tenderly Lift Me: Nurses Honored,
Celebrated, and Remembered*
Jeanne Bryner

7 *The Poetry of Nursing: Poems
and Commentaries of Leading
Nurse-Poets*
Edited by Judy Schaefer

8 *Our Human Hearts: A Medical
and Cultural Journey*
Albert Howard Carter III

9 *Fourteen Stories: Doctors, Patients,
and Other Strangers*
Jay Baruch

10 *Stories of Illness and Healing:
Women Write Their Bodies*
Edited by Sayantani Dasgupta and
Marsha Hurst

11 *Wider than the Sky: Essays and
Meditations on the Healing Power
of Emily Dickinson*
Edited by Cindy Mackenzie and
Barbara Dana

12 *Lisa's Story: The Other Shoe*
Tom Batiuk

13 *Bodies and Barriers: Dramas of
Dis-Ease*
Edited by Angela Belli

14 *The Spirit of the Place: A Novel*
Samuel Shem

15 *Return to* The House of God: *Med-
ical Resident Education 1978–2008*
Edited by Martin Kohn and Carol
Donley

16 *The Heart's Truth: Essays on the
Art of Nursing*
Cortney Davis

17 *Beyond Forgetting: Poetry and
Prose about Alzheimer's Disease*
Edited by Holly J. Hughes

18 *The Country Doctor Revisited: A
Twenty-First Century Reader*
Edited by Therese Zink

19 *The Widows' Handbook: Poetic
Reflections on Grief and Survival*
Edited by Jacqueline Lapidus and
Lise Menn

20 *When the Nurse Becomes a Patient: A Story in Words and Images*
Cortney Davis

21 *What's Left Out*
Jay Baruch

22 *Roses in December: An Alzheimer's
Story*
Tom Batiuk and Chuck Ayers

23 *Mysterious Medicine: The Doctor-
Scientist Tales of Hawthorne and
Poe*
Edited by L. Kerr Dunn

Mysterious Medicine

The Doctor-Scientist Tales of
Hawthorne and Poe

EDITED BY L. KERR DUNN

∾

THE KENT STATE UNIVERSITY PRESS

Kent, Ohio

© 2016 by The Kent State University Press, Kent, Ohio 44242

ALL RIGHTS RESERVED

Library of Congress Catalog Number 2015036093

ISBN 978-1-60635-272-4

Manufactured in the United States of America

LIBRARY OF CONGRESS CATALOGING-IN-PUBLICATION DATA

Mysterious medicine : the doctor-scientist tales of Hawthorne and Poe / edited by
L. Kerr Dunn.

p. ; cm. — (Literature and medicine ; 23)

Includes bibliographical references and index.

ISBN 978-1-60635-272-4 (pbk. : alk. paper) ∞

I. Dunn, L. Kerr, 1971– , editor. II. Series: Literature and medicine (Kent, Ohio) ; 23.

[DNLM: 1. Hawthorne, Nathaniel, 1804–1864. 2. Poe, Edgar Allan, 1809–1849.

3. Medicine in Literature—United States. 4. History of Medicine—United States.

5. History, 19th Century—United States. 6. Science—history—United States. WZ 330]

R149

610.9'034—dc23

2015036093

20 19 18 17 16   5 4 3 2 1

For Jack and Zak, men of mystery and magic.

Contents

⁓

Acknowledgments	ix
A Note on the Texts	xi
Introduction	1
Nathaniel Hawthorne	
Dr. Bullivant (1831)	15
The Haunted Quack: A Tale of a Canal Boat (1831)	23
The Minister's Black Veil: A Parable (1836)	35
Dr. Heidegger's Experiment (1837)	48
Legends of the Province House III: Lady Eleanore's Mantle (1838)	58
The Birthmark (1843)	71
Egotism; or, The Bosom Serpent (1843)	87
Rappaccini's Daughter (1844)	100
Ethan Brand: A Chapter from an Abortive Romance (1850)	127
Edgar Allan Poe	
Sonnet—To Science (1829)	145
Berenice (1835)	148
The Fall of the House of Usher (1839)	158
The Black Cat (1843)	176
A Tale of the Ragged Mountains (1844)	187
The Premature Burial (1844)	198
Mesmeric Revelation (1844)	212
Some Words with a Mummy (1845)	222
The System of Doctor Tarr and Professor Fether (1845)	238
The Facts in the Case of M. Valdemar (1845)	255
Notes	265
References	270

Acknowledgments

⟋⟍

I owe thanks to the following people, who have made this book and my life better: Michael Blackie has been a dream editor. His knowledge, insight, and low-stress approach were invaluable. Joyce Harrison believed in this project and always made me feel welcome as a new Kent State University Press author. Erin Holman was a thoughtful and conscientious copyeditor with whom it was a pleasure to work. Mary Young, Susan Cash, Christine Brooks, and their staff took great care of me throughout the publication process.

My colleagues from the Medical University of South Carolina Writing Center have helped me carve out my professional niche in the health humanities and supported my personal goals as a writer.

Scott Peeples took time to chat with me about Poe one afternoon, giving me renewed excitement about the project as I crossed the finish line. I am indebted to my friend Kathy Beres Rogers for making our meeting possible. She and my other Humanities Committee friends, especially our book club planning committee members, have given me a health humanities community of my own in Charleston over the last six or seven years.

My grad school friends provided me a community of writers at a crucial stage in my writing life—and have continued to support me and to inspire me by their own work.

My parents never doubted that I would grow up to write and have worked every angle to help me be a success, including following famous writers into bathrooms to tell them about their daughter's book. Thanks to all the Kerrs, Dunns, and Littles for years of support and love.

My husband is my best friend, my sounding board, and my anchor. Our son Zak is the light of my life. Thank you both for finding me.

A Note on the Texts

∾

The original publication dates are provided for each tale, poem, or excerpt. The texts of most of Nathaniel Hawthorne's tales are from *The Complete Works of Nathaniel Hawthorne,* edited by George Parsons Lathrop. The following are exceptions: The text of "Dr. Bullivant" is from the 1876 James Osgood & Company Illustrated Library Edition of *Fanshawe, The Dolliver Romance and Other Pieces,* also edited by George Parsons Lathrop. "The Haunted Quack" is from *The Complete Writings of Nathaniel Hawthorne,* edited by Horace E. Scudder and Rose Hawthorne Lathrop for Houghton Mifflin. The texts of Edgar Allan Poe's tales and the poem "Sonnet—To Science" are from *The Complete Works of Edgar Allan Poe,* edited by James Albert Harrison.

Introduction

When as twenty-first-century Americans we say we are "going to the doctor," we typically mean we are going to see a recognized expert in medical care, one who has matriculated at an accredited medical school, is board-certified, and has completed the requisite years of resident training in a select specialty. In some cases, we might see a physician assistant or a nurse practitioner instead of "the doctor"; these clinicians have also completed years of didactic and clinical training in their respective fields, have passed their own certifying exams, and adhere to strict national guidelines that govern practice. Occasionally, we might self-diagnose after watching a television "expert" on an episode of *Dr. Oz, The Doctors, The Dr. Phil Show,* or *Sanjay Gupta, M.D.* The holistic brand of health and healing advocated by Dr. Andrew Weil appeals to many individuals, but his naturopathic approach is still considered nontraditional. In most cases, only when traditional medicine fails does the average American consult practitioners of alternative medicine for serious illness; the term "alternative" itself reflects this bias. Even osteopathic physicians, trained as rigorously as allopathic physicians, face scrutiny—*Exactly how are you different from a regular doctor?* Americans have deeply ingrained ideas about what is standard medical care and what is not.

Our development of this bias did not happen accidentally, but it did happen gradually. The first degreed physicians began to practice in America during the late colonial period, and by the late 1700s the first medical college opened at the University of Pennsylvania (1765) and the Massachusetts Medical Society was formed (1781). But the establishment of these institutions did not mean that Americans started "going to the doctor" in droves. Family members often still acted as primary caregivers (Furst 2000), and when they needed additional help, they looked to various types of folk healers (Browner 2005), both men and women, many who trained through an informal apprentice system. White and black women served as midwives in their respective communities, and on southern plantations African folk healers of both genders cared for other slaves (Cassedy

1991, 51). By the mid-nineteenth century, however, a group of practitioners—primarily white European males, many from well-established families—had begun to centralize power through a number of means, including establishing the American Medical Association in 1847 (Furst 2000; Cassedy 1991; Browner 2005), and to create medical schools that taught a new, "anatomical-clinical" (Long 1989, 22) version of medicine that eventually became the basis of today's mainstream health care.

Yet, even then, the road to regulated medicine was neither smooth nor straightforward. Allopathic practitioners, or "regular" doctors, competed with folk healers, or "irregulars." If you sought health care in the early to mid-nineteenth century, you were, in fact, just as likely to have a healer feel your skull, apply leeches to your flesh, put you into a trance, hand you a bag of herbs, administer a water cure, or urge you to cleanse your spirit to heal your body, as to receive care from a mainstream professional. And so-called mainstreamers often incorporated sectarian elements into their otherwise orthodox practices—or clung to morality-based models of illness left over from the Puritans (Cassedy 1991). Early in the nineteenth century, for example, many regular physicians practiced homeopathy, but by the 1840s they had begun to lose professional clout when physicians like Hawthorne's friend Oliver Wendell Holmes criticized their practices; not until the 1870s, however, did the AMA officially expel homeopathic practitioners from its membership (Browner 2005).[1] In short, for the better part of the nineteenth century, the term "doctor" did not necessarily connote what it does today, for even physicians who did receive more formal professional training were educated in a "haphazard" and "largely unregulated" system (Wendland 2006, 32–33).

Because of this lack of centralized medical power, advances in medical science were slow to take hold in early-nineteenth-century American society. Instead, Americans remained open to a number of experimental, pseudoscientific reforms: mesmerism, which merged science and spirituality to discover truths of human consciousness (Cassedy 1991; Hungerford 1930; Reynolds 2001, 59); homeopathy, which provided alternatives to the allopathic dosing with poisonous medicines like mercury (Stoehr 1978, 106); physiognomy, which aimed at providing a useful way of determining a person's character by analyzing physical features (Benjamin and Baker 2004); and phrenology, which "hoped to provide accurate vocational guidance for the young, to revolutionize our systems of education, to revise the care of the insane, to bring wisdom to the treatment of criminals" by studying the shapes of patients' heads (Hungerford 1930, 210). These reforms, later collectively called pseudoscience, aimed to provide psychological guidance and improve physical health. In this "do-it-yourself" (Stoehr 1978, 271), democratic era, a host of would-be healers and pseudoscientists offered many different paths to care, wellness, and knowledge, some more respectable than others.

One can imagine, then, that even artists and writers felt empowered to contribute to conversations about health and healing. In fact, the American romantics embraced many of the same principles that inform contemporary movements like health humanities and narrative health care. For romantics, truth-seeking required both reason and intuition, science and imagination (Levine and Levine 1990). Today, leaders of the health humanities and narrative movements make a similar case, arguing that arts and humanities research, including narrative health care, should be seen as a complement to evidence-based medicine because it can enrich our understanding of patients' needs and improve their outcomes (Aull 1993; Charon 2006). This rather optimistic viewpoint was reflected in the work of Margaret Fuller, Ralph Waldo Emerson, and Henry David Thoreau, who preached optimism, transcendentalism, and other reforms as paths to better physical, mental, and spiritual health.

Optimism is not, however, characteristic of the dark romantics, like Nathaniel Hawthorne and Edgar Allan Poe, who were concerned with the bleaker side of human nature and wrote about madness, perverseness, self-destruction, and death. And yet, despite a preference for the fantastic, the grotesque, and the macabre, Hawthorne and Poe were deeply engaged in investigating the most significant medical and scientific questions of the day. They exemplified the ways literary narratives can facilitate better understanding of human beings by focusing on *particulars* of character, time, and place (among other factors) that enlarge and even challenge our perceptions of universal truths established through medico-scientific research. The doctor-scientist stories collected here provide evidence that the arts and humanities offer unique methods of exploring the social, cultural, political, and personal forces that affect the way we suffer and heal. They establish Hawthorne and Poe as precursors to the health humanities movement.

On a personal level, Nathaniel Hawthorne and Edgar Allan Poe understood the limits of nineteenth-century medicine well. As a nine-year-old boy, Nathaniel Hawthorne (July 4, 1804–May 19, 1864) injured his foot playing "bat and ball" and was bedridden for fourteen months (Mellow 1980, 18). To move about, he relied on a special support boot and crutches (Wineapple 2001, 15) and did not attend school regularly. He underwent a number of ineffective treatments, including an early version of hydrotherapy that required him to douse his foot in cold water each morning (Mellow 1980, 19; Wineapple 2003). Despite a seemingly miraculous improvement in the winter of 1815, he relapsed that same year. Because of his extended boyhood illness and the related isolation, Hawthorne became an insatiable reader and developed a sense that he was different, a "marked" creature (Wineapple 2001, 15).

Illness and loss were pervasive throughout Hawthorne's life—his father, a sailor, died of yellow fever when Hawthorne was very young, and his mother was often plagued by headaches and other debilitating illnesses (Wineapple 2001, 18). He later proposed marriage to Sophia Peabody, a semi-invalid artist who, much like his mother, suffered from migraines and underwent a number of experimental medical procedures (Mellow 1980, 7). After Hawthorne and Sophia married, Sophia's health improved for a time, and the couple embarked on a happy wedded life, eventually raising three children together. Still, Hawthorne never lost the sense of otherness that would characterize him and his work (Wineapple 2001). If this otherness marked his mind and spirit, it also marked his body: all his life, he would walk with an unbalanced gait. When he grew ill in the early 1860s, he refused to see a doctor, despite his deteriorating condition. Physician Oliver Wendell Holmes, a friend, suspected some "internal organic—perhaps malignant" force at work, but Holmes was not given time to reach a definitive diagnosis (Lathrop 2009, 7). One theory promotes stomach cancer as a likely suspect (Miller 1992).

Younger than Hawthorne by more than four years, Edgar Allan Poe (January 19, 1809–October 7, 1849) also had a childhood characterized by loneliness and loss. His birth parents, both actors, contracted tuberculosis and died young. His alcoholic father disappeared when he was a toddler, and his mother, Eliza, died in December 1811 just before Poe's second birthday (Meyers 1992). Orphaned and separated from his older brother and younger sister, Poe was taken in as the foster child of John and Fanny Allan. Despite the opportunities his foster parents afforded him, Poe was never legally adopted, and he felt like an orphan his whole life, a sense that his rocky relationship with his foster father compounded. As an adult, Poe once wrote, "I have many occasional dealings with Adversity, but the want of parental affection has been the heaviest of my trials" (Quinn 1998, 23).

Ever restless and in need of money, Poe moved from city to city, job to job throughout his life. Even after he married, he did not find long-term happiness, as his young bride Virginia Clemm Poe—famously thirteen years old when they married—died of the same disease that killed Poe's mother. After Virginia died, a woman who nursed her, Mary Louise Shew (later Houghton), daughter of physician Lowrey Barney, continued to minister to Poe, who began suffering from a dangerous illness as early as May 1846. Although in December 1846, he seemed to be recovering his health, in February 1847, the month after Virginia's death, Shew was sending medicine to Poe for fevers and checking in on him (Thomas and Jackson 1987, 640, 693–94). In March, she accompanied Poe to see Dr. Valentine Mott of the New York University School of Medicine, where she told the doctor she believed Poe to have a brain lesion, for "he could not bear stimulants or tonics, without producing insanity" (as quoted in Thomas and Jackson 1987, 694).[2] In early May, another doctor, Dr. John W. Francis, diagnosed Poe with heart disease, which Poe did not accept (Thomas and Jackson 1987, 732). Others have suggested Poe had undiagnosed epilepsy or another underlying pathology that was aggravated by his bouts of drinking.

Whatever the source of Poe's illness, Shew provided steadfast nursing, friendship, and support. Poe felt so endeared to her that he published a tribute, "To M.L.S—," in the *Home Journal* shortly after Virginia's death; another poem, "The Beloved Physician" (also called "The Beautiful Physician"), was not published during Poe's lifetime—although, as Mabbott notes, he at one time allegedly prepared a nine-stanza version for publication (Poe 1978). Shew claimed to have possessed and then lost a manuscript of the poem, and she recorded fragments of it from memory years later for biographer John H. Ingram.[3] Although incomplete and recorded so much later from memory, the poem, along with others written to Shew, offers evidence of Poe's indebtedness to this health care provider.

Despite the medical attention Poe received from Shew and others, the true source of the illness that plagued him remains largely a mystery. His death at the age of forty has been attributed to everything from alcoholism, tuberculosis, or rabies to encephalitis, diabetes, or carbon monoxide poisoning (Peeples 2003). In one of the most in-depth studies of Poe's last days, John Evangelist Walsh (2000) proposed that Poe's death was the result of foul play. As Walsh notes, even the young physician who attended Poe's death, John J. Moran, offered little help in clarifying the exact nature of Poe's death and in later years even contributed to the confusion surrounding Poe's final sickness, as did Poe's friend Dr. Joseph Evans Snodgrass.[4] When Poe's body was exhumed twenty-six years after his death, a small mass that could theoretically have been a calcified tumor was found inside his skull, but this finding remains untested (Geiling 2014).

Poe and Hawthorne's experiences, of course, would have been enough to spark their interest in learning more about health and healing. Certainly a host of scholars have conducted biographical and psychoanalytical readings of Poe and Hawthorne's works that suggest their writings reflect psychological traumas, desires, and motivations.[5] But Hawthorne and Poe each felt an intellectual pull to medical and scientific subjects as well. At Bowdoin College, Hawthorne demonstrated an early interest in medical subjects, enrolling in courses at the newly opened Maine Medical School, where he studied anatomy and physiology

(Wineapple 2003). At West Point, Poe studied mechanical engineering and had access to the "best scientific education available in his country" (Tresch 2002, 118). Outside of formal education, the booming publishing industry offered unprecedented access to a wide variety of medical books like those by Dr. Benjamin Rush and Robley Dunglison (Sloane 1966), self-help manuals, temperance tracts, theological writings, and newspaper articles of "factual" medical accounts, all of which influenced how Americans perceived health and healing.[6]

As readers, writers, and editors, Poe and Hawthorne were more immersed in the world of publishing than the average American. The magazines, books, and papers they read and edited were full of mysterious medical case studies, odd human behavior, and true-life horrors that provided direct inspiration for their tales, including cases where the limitations of medicine resulted in horrific results. Premature burial, for example, was commonly discussed in the nineteenth century, because doctors sometimes mistook comatose or unconscious patients for dead (Dibble 2010); Poe capitalized on the fear of live burial in tales like "The Premature Burial," "Berenice," "The Fall of the House of Usher," and his other beautiful-dead-woman tales, "Ligeia," "Morella," and "Eleonora." Likewise, in "The Masque of the Red Death" he tapped into Americans' fear of epidemics like cholera and tuberculosis that swept through major cities (Levine and Levine 1990). Hawthorne also engaged the theme of epidemics in "Lady Eleanore's Mantle" and the children's tale "The Rejected Blessing" from *Grandfather's Chair,* while simultaneously engaging his interest in American history. Of contemporary sources, Hawthorne had a special fondness for the penny papers and for medical accounts that had sensational appeal, like those of "bosom serpents"—creatures that inhabited the breasts of human beings—that appeared in widely circulated publications, inspiring him to write "Egotism; or, The Bosom Serpent" (Bush 1971).

In particular, Poe and Hawthorne were inspired by the controversies surrounding the various reform movements because these reforms tapped into the philosophical questions about pathologies of the human mind, body, and spirit to which both writers were drawn. However, the two writers were sometimes philosophically opposed, depending on the particular reform in question. While Hawthorne seemed merely amused by phrenology, for example, Poe, in 1836, proclaimed phrenology had achieved the "majesty of a science" (Poe 1984, 252).[7] And where Poe highlighted both the positive and the negative aspects of mesmerism, Hawthorne was repelled by it, viewing it as a violation of the individual soul (Reynolds 1988).[8] When his wife (then fiancée) Sophia—who came from a family of reformers and whose father practiced as a dentist, doctor, and homeopath—received animal magnetism sessions to relieve her crippling headaches, Hawthorne firmly warned her against mesmerism (Stoehr 1978, 42–48). Ultimately, both Poe and Hawthorne regarded pseudoscience with a mixture of skepticism and wonder.

Despite their ambivalent attitudes toward reforms—or perhaps because of them—Poe and Hawthorne both incorporated elements of phrenology and physiognomy in numerous tales and essays, wrote major works that included mesmerism, and felt free to criticize reformers when they sensed fraud or improvidence. These reforms—along with other movements in spiritualism and mental healing—were the prominent practices in psychology before the arrival of the new, science-based psychology late in the century (Benjamin and Baker 2004). And Poe and Hawthorne were keenly interested in human psychology.

Elements of phrenology and physiognomy are embedded in a number of Poe tales, including "The Fall of the House of Usher," in which Poe uses multiple allusions to characterize Roderick Usher (Zimmerman 2007), and "The Imp of the Perverse," in which the narrator describes the perverseness in human nature that he claims phrenologists had yet to explain. In "The System of Doctor Tarr and Professor Fether," Poe mocks the "moral treatment" or "soothing system" that became a popular mental health reform based on the work of French physician Philippe Pinel (Benton 1968, 7; Benjamin and Baker 2004). In two of his mesmerism tales, "A Tale of the Ragged Mountains" and "Mesmeric Revelation," Poe explores the talents of mesmerist-doctors who have forged strong rapports with their patients, relationships that bring the patients physical comfort and spiritual awareness. But in his third mesmerism tale, "The Facts in the Case of M. Valdemar," Poe portrays mesmerism as an eerie power that holds sway over the individual spirit, illustrating the potentially gruesome and inexplicable consequences of experimenting on human subjects.

Like Poe, Hawthorne uses phrenology and physiognomy for literary effect in early tales, including "Roger Malvin's Burial" and "My Kinsman, Major Mollineaux." Yet, though he had been exposed to phrenology at Bowdoin while attending a special lecture series taught by phrenology proponent Dr. John Doane Wells (Stoehr 1978), Hawthorne was no convert. For example, he mocks phrenology in his 1828 novel *Fanshawe* and a decade later in the satirical "The Science of Noses" (Stoehr 1974), which he wrote as editor of the *American Magazine for Useful and Entertaining Knowledge*. Throughout his career, he continued to demonstrate suspicion of the various pseudosciences, as evidenced in his later romances and unfinished works, "Septimius Felton" and "The Dolliver Romance."

But Hawthorne was not merely critical of pseudoscientists. He was also wary of the new clinical medicine imported from the French clinics that focused on "pathological anatomy" (Foucault 1994, 124). As American physicians began to import new medical practices from the French clinics, Hawthorne and Poe took note of how this new form of medicine altered perceptions of human mental and physical illness. Physicians who practiced it placed emphasis on the empirical study of the body (Furst 2000; Goldman 2004). Although this focus allowed for better understanding of disease processes, it also energized a number of unsavory practices, such as grave robbing and unethical experimentation on members of disenfranchised groups like women and slaves. Some doctors and scientists looked to the body, for example, to find physical evidence of these groups' biological inferiority (Savitt 1997, Welter 1966). While Hawthorne and Poe were by no means politically active in the women's rights or abolitionist movements, like other American romantics, they were attuned to the potential for abuses of power in medicine and science. Certainly, they both described how an obsessive fixation on the body could have dire consequences. In tales like "Berenice," "The Birthmark," and others, Hawthorne and Poe portray characters with monomaniacal fixations on the body—the female body in particular—demonstrating how a way of gazing at another human being could become pathological and posing questions about the connections among mind, body, and spirit that the new empirically driven medicine largely dismissed.

With this new way of seeing the body and disease also came a new way of talking about the body and disease. Professional jargon often serves as a dividing line between expert and nonexpert, but Poe mastered the language of clinical medicine so well that doctors debated the legitimacy of his mesmerism tales as if they were actual medical case reports (Thomas and Jackson 1987). Poe's awareness of the power of health rhetoric also informs other doctor-scientist tales. For example, in "The System of Doctor Tarr and Professor Fether," patients in a mental asylum learn how to mimic their doctor's "psychobabble" (Long 1989, 23), a key ingredient of their usurpation of power. In "The Fall of the House of Usher," the family physician is struck speechless, and his departure from the house empowers Roderick Usher and the narrator to attempt their own diagnoses (Hardy 1998). These novices' abilities to exercise power in medical scenarios, where professionals fail, reflect the "unprecedented flexibility and creativity" of the reform spirit (Reynolds 1988, 243): Americans as a whole were a willing audience for anyone who could walk the walk and talk the talk.

And whereas many of Poe's characters fail as diagnosticians, Poe had a talent for describing characters with a clinical accuracy that makes him seem somewhat prescient (Hardy 1998; Zimmerman 2001). Literary scholar Brett Zimmerman (2001) has argued, for example, that Poe provides a case study of paranoid schizophrenia in "The Tell-Tale Heart" decades before schizophrenia was named or understood as a discrete mental illness.[9] Altschuler and Augenstein (2012) note that Peter Pendulum in "The Businessman" exhibits signs of frontal lobe syndrome before the famous case of Phineas Gage put the syndrome on the map. Poe's visionary ideas are also reflected in "The Imp of the Perverse," in which the narrator points out the failure of both phrenologists and moralists to understand human behavior, anticipating later theories about the subconscious mind and repression (Hoffman 1998). Finally, in "The Man Who Was Used Up," Poe describes advances in the science of bionics and prosthetics, touching upon themes related to the mind, body, and identity that seem ahead of his time.

"A Tale of the Ragged Mountains" represents this type of prescient-diagnostic tale, as physician Robert Battle (2011) suggests that the character Bedloe is "a profound phenotype of Marfan syndrome" (148). Although he was not necessarily prescient in his understanding of alcoholism, in "The Black Cat" Poe offers a nuanced and subversive take on the temperance tracts of his time that draws upon the then newly emerging concept of an alcoholic not as a person with weak moral character but a person with a disease (Matheson 1986). Whether or not we agree that Poe seems "positively clairvoyant" (Zimmerman 2009, 7), he does exhibit a keen ability to merge scientific knowledge with intuition for the purpose of creating portraits of mentally and physically sick individuals whose stories remind us that while science can teach us how a disease affects a body, narratives, nonfiction and fiction, can teach us how a person experiences an illness. As a result, his tales demonstrate that Poe was correct to insist on the need for "alternatives to rational, empirical ways of knowing" (Peeples 1998, 28)—a romantic theme that was introduced in his early poem "Sonnet—To Science" and continued to underpin his entire body of work.

If Poe became an informal student of medicine, drawing on the available texts of his time the way later writers would draw upon the works of Freud and Jung (Zimmerman 1992), the same might be said of Hawthorne, whose reading life included not only forays into travel, history, and crime narratives but also scientific and medical materials from books, newspapers, and magazines (Reynolds 1988; Bush 1971; Browner 2005). His medical-themed tales reveal an impressive knowledge of medical history (Dolezal 2005) and are informed by a variety of sources from literature to medical case histories to theological writings (Bush 1971; Turner 1936). Even Hawthorne's personal observations reflect his fascination for the doctor figure. Both his *French and Italian Notebooks* and his *American Notebooks* include descriptions of doctors he observed in real life or imagined creating, and he even claimed to have been haunted by the ghost of a doctor while reading in the Boston Athenaeum, an experience he describes in an autobiographical piece, "The Ghost of Doctor Harris."[10]

In his tales, Hawthorne translated source material into fictional profiles of individuals afflicted by physical or mental pathologies, and those afflicted individuals were often the doctors themselves. While in Poe's illness-themed tales doctors are frequently only nominally present, Hawthorne often places them at the heart of the story, portraying—with a few notable exceptions, like Doctor Clarke in "Lady Eleanore's Mantle"—doctor-scientists with hubristic aims. Even minor doctor characters, like the one in "The Great Carbuncle," usually fit this type, although in historically inspired pieces like "Dr. Bullivant" and "The Rejected Blessing" Hawthorne is more concerned with the social and political underpinnings of the medical world. His historical tales, however, are influenced by nineteenth-century preoccupations with pseudoscientific reforms and experimental practices; as a result, Hawthorne's doctor-scientist characters often reflect not one school of thought or time period but a variety of approaches from past, present, and even future.

In two of Hawthorne's earliest works, "Dr. Bullivant" and "The Haunted Quack," the doctor-scientist is a peddler of various herbs and elixirs. Dr. Bullivant is an apothecary whose political activities put a rift between him and his community, while the eponymous doctor in "The Haunted Quack" is a poorly trained, monetarily motivated physician who is haunted by a patient he believes he has killed. The danger of elixirs and other pharmaceuticals is a consistent motif in Hawthorne's tales, one that highlights the mix of magic and detachment that for Hawthorne characterized medicine. In "Dr. Heidegger's Experiment," "Rappaccini's Daughter," and "The Birthmark," doctor-scientists practice enigmatic and, literally, poisonous brands of alchemy that destroy their patients, whose health the doctor-scientists had made secondary to scientific ambitions.

If their attachments to magical potions and unethical experimentation link Hawthorne's doctors to early-century pseudoscientists of whom Hawthorne was suspicious, then their lack of empathy and their obsessions with the physical body link them to the new clinical medicine (Long 1989; Browner 1993). In "Egotism: or, The Bosom Serpent" and "Ethan Brand," Hawthorne juxtaposes the new clinical medicine that sought a physical source for all illness, even mental, with the religious belief that all illness originated in moral failing (Goldman 2004). In doing so, Hawthorne underscores the reductive and, therefore, ineffectual approaches of each (Goldman 2004), while simultaneously highlighting the similar ways both groups envision sickness as pathology written on the body. This tension between and conflation of old and new theories of medicine is further illustrated in his masterpiece, *The Scarlet Letter,* in which Hawthorne depicts physician Roger Chillingworth as a "medical palimpsest" (Dolezal 2005, 17), a character who not only combines ancient practices with a mid-century method of clinical observation but anticipates future psychoanalytical techniques (Stoehr 1978; Browner 1993). Although the novel is set in Puritan New England, Hawthorne brings to it a nineteenth-century sensibility that allows him to trace the history of medicine while anticipating its future incarnations.

Hawthorne's final romances, "Dr. Grimshawe's Secret," "Septimius Felton," and "The Dolliver Romance," all of which he left uncompleted and unpublished at the time of his death, also invite us to compare past, present, and future. In "The Dolliver Romance," for example, we find Hawthorne still intrigued by the supernatural but also exploring sociohistorical themes related to medicine. In this romance, Hawthorne alludes to the dual influences of Aesculapius, Greek demigod of medicine and healing, and Hippocrates, the Greek physician who separated medicine from religion and gave it a more scientific basis. In his descriptions of the professional rivalry between formally and informally trained professionals, Hawthorne provides evidence that the early American medical profession was defined by conflicting influences and practices. Thus, even at the end of his career, Hawthorne demonstrated an ongoing and perhaps refined understanding of the evolution of medicine and science in America. For both authors, the mysteries of the human heart, mind, and body along with their ailments were worth visiting and revisiting, as if each unique story could add an inimitable layer to our collective understanding of sickness and suffering.

Hawthorne and Poe wrote numerous works that relate to the medical themes of this anthology. Knowing that I cannot collect in this slender volume all illness-themed works produced by Hawthorne and Poe, I focused primarily on tales that feature a doctor-scientist, even when the appearance of that character is quite brief. For example, although the physicians in Poe's "Berenice" and "The Fall of the House of Usher" and Hawthorne's "The Minister's Black Veil" and "Ethan Brand" appear only briefly (in "Berenice," the family physician is merely mentioned and does not appear), those characters deserve closer study through the lens of health humanities.[11] At the same time, I did not include Hawthorne's "The Great Carbuncle," because the doctor is a minor character whose type is realized more fully in other tales, or "The Rejected Blessing," a historically themed children's tale that is, despite my omission, worthy of study in the context of health humanities, as Cerulli and Berry (2014) have demonstrated. In contrast, although it does not feature a doctor-scientist character, I include Poe's "The Black Cat" because it introduces Poe's concept of "perverseness," an idea that shows Poe developing his own theories of illness. While Poe's tale "The Imp of the Perverse" arguably tackles this topic of perverseness more directly, I selected "The Black Cat" instead because it offered the added opportunity to analyze nineteenth-century attitudes toward alcoholism.

In addition to the tales, I collect one poem from Poe and one historical portrait from Hawthorne. Poe's "Sonnet—To Science" is included because its theme conveys an early version of the romantic philosophy that guided much of his work that is relevant to the health humanities. Hawthorne's sketch, "Dr. Bullivant," represents one of his first published works about the doctor-apothecary, a figure that bookends his career, as Hawthorne returned to it in his final, uncompleted work, "The Dolliver Romance." The sketch also reflects Hawthorne's interest in social history, which would inform his creation of doctor-scientists. While I have not included excerpts from Hawthorne's novels, I urge readers and students to read them, complete and unfinished, because they, too, feature doctors and pseudoscientists worthy of study in the context of health humanities.

Collected here are works that invite us to compare historical and contemporary depictions of healers and patients. They call us to consider our own perceptions of "going to the doctor" in an increasingly complex interprofessional health care system, a system defined more often in terms of team-based and relationship-centered care. The stories represent creativity as inquiry, unsettling our typical notions of science and medicine even as they address familiar topics. They speak to us about medical ethics and experimentation, the perils of hubristic practice and abuse of power, patient-provider relationships, aging and the promise of eternal youth, quests for perfection and immortality, the effects of epidemics on society, the psychological processes of the mentally ill, and the relationship between the body and identity. Although written nearly two centuries ago, these works illuminate contemporary issues related to medical error, alternative therapies, plastic surgery, cloning, genetic profiling, mental health stigmas, health care disparities and inequalities, women's health, and end-of-life issues. They call us to analyze what it means to live in a time of rapid change—and to remember what happens when we abandon humanistic principles in the name of progress. They remind us that we, too, live in times of great uncertainty that require careful reflection, insight, and imagination, even skepticism. They suggest that even in fantasy we might examine critically issues that affect our perceptions of health at both the individual and the cultural levels. So we enter the fantasies into which Poe and Hawthorne invite us—drink the magic elixir, pry open the coffin, submit ourselves to be mesmerized. And in fantasy we discover truths otherwise obscured; we remember that to see our world clearly, we must view it through not one lens but many.

Nathaniel Hawthorne

❧

Dr. Bullivant

January 1831, *Salem Gazette*

Nathaniel Hawthorne's interest in medicine and science did not develop gradually; it was strong from the start of his writing career. About five years after being graduated as a member of the famous 1825 class of Bowdoin College, where he took anatomy and physiology classes at the newly established Maine Medical School, Hawthorne published two works that featured doctors as main characters: the unsigned sketch "Dr. Bullivant" published in the Salem Gazette *and the tale "The Haunted Quack," attributed to Hawthorne although published in the* Token *under the pseudonym Joseph Nicholson. While these pieces are considered minor works in the Hawthorne canon, in the context of the health humanities they offer a significant introduction to the medico-scientific themes and characters Hawthorne would develop in many of his more recognizable works.*

In "Dr. Bullivant," Hawthorne presents a vivid, multidimensional portrait of an apothecary practicing in 1670 New England. While conducting research on New England history in books like John Dunton's Life and Errors, *which describes Salem and Boston in 1686, Hawthorne likely selected Bullivant as a figure who reflected "the national character" as well as "broad human and moral problems" (Turner 1980, 56). His descriptions of Bullivant's apothecary shop, easily identified by the bust of Aesculapius, Greek demigod of medicine and healing, provide insights into the role of the prerevolutionary apothecary. At the same time, Hawthorne offers a snapshot of colonial politics that leave Dr. Bullivant, at least temporarily, on the wrong side of American history. A man of "medicine and politics," Bullivant is a controversial character but a firm fixture in his community nonetheless. Arguably, his neighbors play a significant role both in his demise and in his recovery, making this story suitable for discussions about the role community plays in supporting and promoting the health of its members. Because Bullivant dispenses jokes along with his numerous medicines, the sketch also offers unique opportunities to discuss the role of humor in health care.*

Readers may also enjoy comparing Dr. Bullivant to Dr. Dolliver in "The Dolliver Romance," one of Hawthorne's last doctors, who is also an apothecary.

Additionally, readers may be inspired to trace the evolution of the apothecary from colonial times to the present, when physicians and pharmacists train in distinct specialties. Finally, as it was common courtesy to address an apothecary as "doctor" in colonial times, health humanities students might discuss social and professional rules that govern the use of the title "doctor" in an increasingly complex and interprofessional system in which various health care professionals other than physicians also earn doctorate-level degrees.

•

THIS person was not eminent enough, either by nature or circumstance, to deserve a public memorial simply for his own sake, after the lapse of a century and a half from the era in which he flourished. His character, in the view which we propose to take of it, may give a species of distinctness, and point to some remarks on the tone and composition of New England society, modified as it became by new ingredients from the eastern world, and by the attrition of sixty or seventy years over the rugged peculiarities of the original settlers. We are perhaps accustomed to employ too sombre a pencil in picturing the earlier times among the Puritans, because at our cold distance, we form our ideas almost wholly from their severest features. It is like gazing on some scenes in the land which we inherit from them; we see the mountains, rising sternly and with frozen summits up to heaven, and the forests, waving in massy depths where sunshine seems a profanation, and we see the gray mist, like the duskiness of years, shedding a chill obscurity over the whole; but the green and pleasant spots in the hollow of the hills, the warm places in the heart of what looks desolate, are hidden from our eyes. Still, however, a prevailing characteristic of the age was gloom, or something which cannot be more accurately expressed than by that term, and its long shadow, falling over all the intervening years, is visible, though not too distinctly, upon ourselves. Without material detriment to a deep and solid happiness, the frolic of the mind was so habitually chastened, that persons have gained a nook in history by the mere possession of animal spirits, too exuberant to be confined within the established bounds. Every vain jest and unprofitable word was deemed an item in the account of criminality, and whatever wit, or semblance thereof, came into existence, its birthplace was generally the pulpit, and its parent some sour old Genevan divine. The specimens of humor and satire, preserved in the sermons and controversial tracts of those days, are occasionally the apt expressions of pungent thoughts; but oftener they are cruel torturings and twistings of trite ideas, disgusting by the wearisome ingenuity which constitutes their only merit. Among a people where so few possessed, or were allowed to exercise, the art of extracting the mirth which lies hidden like latent caloric in almost everything, a gay apothecary, such as Dr. Bullivant, must have been a phenomenon.

We will suppose ourselves standing in Cornhill, on a pleasant morning of the year 1670, about the hour when the shutters are unclosed, and the dust swept from the doorsteps, and when Business rubs its eyes, and begins to plod sleepily through the town. The street, instead of running between lofty and continuous piles of brick, is but partially lined with wooden buildings of various heights and architecture, in each of which the mercantile department is connected with the domicile, like the gingerbread and candy shops of an after-date. The signs have a singular appearance to a stranger's eye. These are not a barren record of names and occupations, yellow letters on black boards, but images and hieroglyphics, sometimes typifying the principal commodity offered for sale, though generally intended to give an arbitrary designation to the establishment. Overlooking the bearded Saracens, the Indian Queens, and the wooden Bibles, let us direct our attention to the white post newly erected at the corner of the street, and surmounted by a gilded countenance which flashes in the early sunbeams like veritable gold. It is a bust of Æsculapius, evidently of the latest London manufacture; and from the door behind it steams forth a mingled smell of musk and assafœtida and other drugs of potent perfume, as if an appropriate sacrifice were just laid upon the altar of the medical deity. Five or six idle people are already collected, peeping curiously in at the glittering array of gallipots and phials, and deciphering the labels which tell their contents in the mysterious and imposing nomenclature of ancient physic. They are next attracted by the printed advertisement of a Panacea, promising life but one day short of eternity, and youth and health commensurate. An old man, his head as white as snow, totters in with a hasty clattering of his staff, and becomes the earliest purchaser, hoping that his wrinkles will disappear more swiftly than they gathered. The Doctor (so styled by courtesy) shows the upper half of his person behind the counter, and appears to be a slender and rather tall man; his features are difficult to describe, possessing nothing peculiar, except a flexibility to assume all characters in turn, while his eye, shrewd, quick, and saucy, remains the same throughout. Whenever a customer enters the shop, if he desire a box of pills, he receives with them an equal number of hard, round, dry jokes,—or if a dose of salts, it is mingled with a portion of the salt of Attica,—or if some hot, Oriental drug, it is accompanied by a racy word or two that tingle on the mental palate,—all without the least additional cost. Then there are twistings of mouths which never lost their gravity before. As each purchaser retires, the spectators see a resemblance of his visage pass over that of the apothecary, in which all the ludicrous points are made most prominent, as if a magic looking-glass had caught the reflection, and were making sport with it. Unwonted titterings arise and strengthen into bashful laughter, but are suddenly hushed as some minister, heavy-eyed from his last night's vigil, or magistrate, armed with the terror of the whipping-post and pillory, or perhaps the governor himself, goes by like a dark cloud intercepting the sunshine.

About this period, many causes began to produce an important change on and beneath the surface of Colonial society. The early settlers were able to keep within the narrowest limits of their rigid principles, because they had adopted them in mature life, and from their own deep conviction, and were strengthened in them by that species of enthusiasm, which is as sober and as enduring as reason itself. But if their immediate successors followed the same line of conduct, they were confined to it, in a great degree, by habits forced upon them, and by the severe rule under which they were educated, and in short more by restraint than by the free exercise of the imagination and understanding. When therefore the old original stock, the men who looked heavenward without a wandering glance to earth, had lost a part of their domestic and public influence, yielding to infirmity or death, a relaxation naturally ensued in their theory and practice of morals and religion, and became more evident with the daily decay of its most strenuous opponents. This gradual but sure operation was assisted by the increasing commercial importance of the colonies, whither a new set of emigrants followed unworthily in the track of the pure-hearted Pilgrims. Gain being now the allurement, and almost the only one, since dissenters no longer dreaded persecution at home, the people of New England could not remain entirely uncontaminated by an extensive intermixture with worldly men. The trade carried on by the colonists (in the face of several inefficient acts of Parliament), with the whole maritime world, must have had a similar tendency; nor are the desperate and dissolute visitants of the country to be forgotten among the agents of a moral revolution. Freebooters from the West Indies and the Spanish Main,—state criminals, implicated in the numerous plots and conspiracies of the period,—felons, loaded with private guilt,—numbers of these took refuge in the provinces, where the authority of the English king was obstructed by a zealous spirit of independence, and where a boundless wilderness enabled them to defy pursuit. Thus the new population, temporary and permanent, was exceedingly unlike the old, and far more apt to disseminate their own principles than to imbibe those of the Puritans. All circumstances unfavorable to virtue acquired double strength by the licentious reign of Charles II.; though perhaps the example of the monarch and nobility was less likely to recommend vice to the people of New England than to those of any other part of the British Empire.

The clergy and the elder magistrates manifested a quick sensibility to the decline of godliness, their apprehensions being sharpened in this particular no less by a holy zeal than because their credit and influence were intimately connected with the primitive character of the country. A Synod, convened in the year 1679, gave its opinion that the iniquity of the times had drawn down judgments from Heaven, and proposed methods to assuage the Divine wrath by a renewal of former sanctity. But neither the increased numbers nor the altered spirit of the people, nor the just sense of a freedom to do wrong, within certain limits, would now have permitted the exercise of that inquisitorial strictness, which had been wont to penetrate to men's firesides and watch their domestic life, recognizing no distinction between private ill conduct and crimes that endanger the community. Accordingly, the tide of worldly principles encroached more and more upon the ancient landmarks, hitherto esteemed the outer boundaries of virtue. Society arranged itself into two classes, marked by strong shades of difference, though separated by an uncertain line;—in one were included the small and feeble remnant of the first settlers, many of their immediate descendants, the whole body of the clergy, and all whom a gloomy temperament, or tenderness of conscience, or timidity of thought, kept up to the strictness of their fathers; the other comprehended the new emigrants, the gay and thoughtless natives, the favorers of Episcopacy, and a various mixture of liberal and enlightened men with most of the evil doers and unprincipled adventurers in the country. A vivid and rather a pleasant idea of New England manners, when this change had become decided, is given in the journal of John Dunton, a cockney bookseller, who visited Boston and other towns of Massachusetts with a cargo of pious publications, suited to the Puritan market. Making due allowance for the flippancy of the writer, which may have given a livelier tone to his descriptions than truth precisely warrants, and also for his character, which led him chiefly among the gayer inhabitants, there still seems to have been many who loved the winecup and the song, and all sorts of delightful naughtiness. But the degeneracy of the times had made far less progress in the interior of the country than in the seaports, and until the people lost the elective privilege, they continued the government in the hands of those upright old men who had so long possessed their confidence. Uncontrollable events, alone, gave a temporary ascendency to persons of another stamp. James II., during the four years of his despotic reign, revoked the charters of the American colonies, arrogated the appointment of their magistrates, and annulled all those legal and prescriptive rights which had hitherto constituted them nearly independent states. Among the foremost advocates of the royal usurpations was Dr. Bullivant. Gifted with a smart and ready intellect, busy and bold, he acquired great influence in the new government, and assisted Sir Edmund Andros, Edward Randolph, and five or six others, to browbeat the council, and misrule the northern provinces according to their pleasure. The strength of the popular hatred against this administration, the actual tyranny that was exercised, and the innumerable fears and jealousies, well grounded and fantastic, which harassed the country, may be best learned from a work of Increase Mather, the "Remarkable Providences of the Earlier Days of American Colonization." The good divine (though writing when a lapse of nearly forty years should have tamed the fierceness of party animosity) speaks with the most bitter and angry scorn of "Pothecary Bullivant," who probably indulged his satirical propensities, from the seat of power, in a manner which rendered him an especial object of public dislike. But the people were about to play off a piece of practical fun on the Doctor and the whole of his coadjutors, and have the laugh all to themselves. By the first faint rumor of the attempt of the Prince of Orange on the throne, the power of James was annihilated in the colonies, and long before the abdication of the latter became known, Sir Edmund Andros, Governor-General of New England and New York, and fifty of the most obnoxious leaders of the court party, were tenants of a prison. We will visit our old acquaintance in his adversity.

The scene now represents a room of ten feet square, the floor of which is sunk a yard or two below the level of the ground; the walls are covered with a dirty and crumbling plaster, on which appear a crowd of ill-favored and lugubrious faces done in charcoal, and the autographs and poetical attempts of a long succession of debtors and petty criminals. Other features of the apartment are a deep fireplace (superfluous in the sultriness of the summer's day), a door of hard-hearted oak, and a narrow window high in the wall, where the glass has long been broken, while the iron bars retain all their original strength. Through this opening come the sound of passing footsteps in the public street, and the voices of children at play. The furniture consists of a bed, or rather an old sack of barley straw, thrown down in the corner farthest from the door, and a chair and table, both aged and infirm, and leaning against the side of the room, besides lending a friendly support to each other. The atmosphere is stifled and of an ill smell, as if it had been kept close prisoner for half a century, and had lost all its pure and elastic nature by feeding the tainted breath of the vicious and the sighs of the unfortunate. Such is the present abode of the man of medicine and politics, and his own appearance forms no contrast to the accompaniments. His wig is unpowdered, out of curl, and put on awry; the dust of many weeks has worked its way into the web of his coat and small-clothes, and his knees and elbows peep forth to ask why they are so ill clad; his stockings are ungartered, his shoes down at the heel, his waistcoat is without a button, and discloses a shirt as dingy as the remnant of snow in a showery April day. His shoulders have become rounder, and his whole person is more bent and drawn together, since we last saw him, and his face has exchanged the glory of wit and humor for a sheepish dullness. At intervals, the Doctor walks the room, with an irregular and shuffling pace; anon, he throws himself flat on the sack of barley straw, muttering very reprehensible expressions between his teeth; then again he starts to his feet, and journeying from corner to corner, finally sinks into the chair, forgetful of its three-legged infirmity till it lets him down upon the floor. The grated window, his only medium of intercourse with the world, serves but to admit additional vexations. Every few moments the steps of the passengers are heard to pause, and some well-known face appears in the free sunshine behind the iron bars, brimful of mirth and drollery, the owner whereof stands on tiptoe to tickle poor Dr. Bullivant with a stinging sarcasm. Then laugh the little boys around the prison door, and the wag goes chuckling away. The apothecary would fain retaliate, but all his quips and repartees, and sharp and facetious fancies, once so abundant, seem to have been transferred from himself to the sluggish brains of his enemies. While endeavoring to condense his whole intellect into one venomous point, in readiness for the next assailant, he is interrupted by the entrance of the turnkey with the prison fare of Indian bread and water. With these dainties we leave him.

When the turmoil of the Revolution had subsided, and the authority of William and Mary was fixed on a quiet basis throughout the colonies, the deposed governor and some of his partisans were sent home to the new court, and the others released from imprisonment. The New-Englanders, as a people, are not apt to retain a revengeful sense of injury, and nowhere, perhaps, could a politician, however odious in his power, live more peacefully in his nakedness and disgrace. Dr. Bullivant returned to his former occupation, and spent rather a desirable old age. Though he sometimes hit hard with a jest, yet few thought of taking offence; for whenever a man habitually indulges his tongue at the expense of all his associates, they provide against the common annoyance by tacitly agreeing to consider his sarcasms as null and void. Thus for many years, a gray old man with a stoop in his gait, he continued to sweep out his shop at eight o'clock in summer mornings and nine in the winter, and to waste whole hours in idle talk and irreverent merriment, making it his glory to raise the laughter of silly people, and his delight to sneer at them in his sleeve. At length, one pleasant day, the door and shutters of his establishment kept closed from sunrise till sunset, and his cronies marvelled a moment, and passed on;—a week after, the rector of King's Chapel said the death-rite over Dr. Bullivant;—and within the month a new apothecary, and a new stock of drugs and medicines, made their appearance at the gilded Head of Æsculapius.

DISCUSSION QUESTIONS

1. Doctor Bullivant is described as a man who practices "the art of extracting mirth which lies hidden like latent caloric in almost everything." Analyze Hawthorne's descriptions of the role of humor played in colonial times and the role it played in Bullivant's life. When is the use of humor appropriate in health care? What purpose might it serve for some health care providers? Imagine a scenario in which a provider might use humor inappropriately, especially when communicating with patients and colleagues.

2. The narrator notes that Bullivant is "so styled" a doctor "by courtesy." In an increasingly interprofessional twenty-first century health care system, in which pharmacists, dentists, nurses, and physical therapists can also earn degrees that confer the title "doctor," do you predict the term will continue to evolve? Speculate on how these changes will reflect shifting relationships among health care providers—and what communication needs may arise as a result.

3. How do Bullivant's customers compare to twenty-first-century health care consumers? What do their similarities suggest about the timelessness of certain human attitudes toward health, aging, and beauty?

4. A bust of Aesculapius stands at the front of Bullivant's apothecary shop— and references to it frame the sketch. What was this demigod's association with the apothecary trade, and what metaphorical relevance does this figure have in the sketch? Does Aesculapius—or the values he represented—still have relevance in the world of health care today?

5. What mental or physical health challenges does Bullivant face when he goes to prison? How do these speak to what we might today call the biocultural needs of individuals? What role do community and culture play in Bullivant's recovery of his former identity, occupation, and health?

6. Research and write a history of the apothecary in America and share it with the class or your reading group. In particular, discuss the various roles the apothecary served—from midwifery to surgery to pharmacology—and the implications of splitting those roles into distinct specialties as they exist today.

The Haunted Quack
A Tale of a Canal Boat
1831, *Token*

Quacks were a staple of nineteenth-century America, from the ubiquitous traveling snake-oil salesmen, to the Lydia Pinkhams who mass-marketed home remedies, to the local apothecaries who believed their experimental herbal concoctions offered real comfort—or, at least, allegedly did no harm.[1] In the first half of the century, when practicing medicine required no formal medical training and doctors could offer little in the way of true healing, Americans either relied on their own homegrown remedies or bravely tested the mysterious salves and potions these doctors had to offer. In "The Haunted Quack," Hawthorne provides a somewhat humorous depiction of a young doctor whose uninformed medical practices come back to haunt him.

This story, which has been attributed to Hawthorne, although published under the alleged pseudonym Joseph Nicholson, finds Hawthorne early on in his career highlighting the questionable practices of doctors whose various potions and alchemies often wrought havoc on patients.[2] Although lighter in tone and more sympathetic to doctors than later stories, this tale anticipates Hawthorne's more virulent attacks on unethical medical and scientific experimentation. Yet, while he pokes fun at the young doctor, Hippocrates Jenkins, Hawthorne portrays the actual dilemmas faced by doctors and patients in an era when patients were vulnerable and incidences of medical error—or outright fraud—were high.

The tale exemplifies how the methods of training as well as the attitudes, habits, and ethics of practitioners influence the quality of care delivered by the next generation. Readers might compare to present-day training the apprenticeship system of education described. Clinical teachers might be prompted to share their own stories about practicing with professionalism and recommend strategies for students to keep abreast of the most current knowledge in their field. And, certainly, a discussion of medical ethics with a focus on medical error is warranted—as today's practitioners can be "haunted" by their own mistakes much as the narrator of this story.

•

IN the summer of 18—, I made an excursion to Niagara. At Schenectady, finding
the roads nearly impassable, I took passage in a canal boat for Utica. The weather
was dull and lowering. There were but few passengers on board; and of those
few, none were sufficiently inviting in appearance to induce me to make any
overtures to a travelling acquaintance. A stupid answer, or a surly monosyllable,
were all that I got in return for the few simple questions I hazarded. An occasional
drizzling rain, and the wet and slippery condition of the tow-path, along which
the lazy beasts that dragged the vessel travelled, rendered it impossible to vary
the monotony of the scene by walking. I had neglected to provide myself with
books, and as we crept along at the dull rate of four miles per hour, I soon felt
the foul fiend *Ennui* coming upon me with all her horrors.

"Time and the hour," however, "runs through the roughest day," and night at
length approached. By degrees the passengers, seemingly tired of each other's
company, began to creep slowly away to their berths; most of them fortifying
themselves with a potation, before resigning themselves to the embrace of Mor-
pheus. One called for a glass of hot whiskey punch, because he felt cold; another
took some brandy toddy to prevent his taking cold; some took mint juleps;
some gin-slings, and some rum and water. One took his dram because he felt
sick; another to make him sleep well; and a third because he had nothing else
to do. The last who retired from the cabin, was an old gentleman who had been
deeply engaged in a well-thumbed volume all day, and whose mental abstraction
I had more than once envied. He now laid down his book, and, pulling out a red
nightcap, called for a pint of beer, to take the vapors out of his head.

As soon as he had left the cabin, I took up the volume, and found it to be
Glanville's marvellous book, entitled the History of Witches, or the Wonders
of the Invisible World Displayed. I began to peruse it, and soon got so deeply
interested in some of his wonderful narrations, that the hours slipped uncon-
sciously away, and midnight found me poring half asleep over the pages. From
this dreamy state I was suddenly aroused by a muttering, as of a suppressed
voice, broken by groans and sounds of distress. Upon looking round, I saw
that they proceeded from the figure of a man enveloped in a cloak who was
lying asleep upon one of the benches of the cabin whom I had not previously
noticed. I recognized him to be a young man, with whose singular appearance and behavior during the day, I had been struck. He was tall and thin in
person, rather shabbily dressed, with long, lank, black hair, and large gray
eyes, which gave a visionary character to one of the most pallid and cadaverous countenances I had ever beheld. Since he had come on board, he had

appeared restless and unquiet, keeping away from the table at meal times, and seeming averse from entering into conversation with the passengers. Once or twice, on catching my eye, he had slunk away as if, conscience-smitten by the remembrance of some crime, he dreaded to meet the gaze of a fellow mortal. From this behavior I suspected that he was either a fugitive from justice, or else a little disordered in mind; and had resolved to keep my eye on him and observe what course he should take when we reached Utica.

Supposing that the poor fellow was now under the influence of nightmare, I got up with the intention of giving him a shake to rouse him, when the words, "murder," "poison," and others of extraordinary import, dropping unconnectedly from his lips, induced me to stay my hand. "Go away, go away," exclaimed he, as if conscious of my approach, but mistaking me for another. "Why do you continue to torment me? If I did poison you, I didn't mean to do it, and they can't make that out more than manslaughter. Besides, what's the use of haunting me now? Ain't I going to give myself up, and tell all? Begone! I say, you bloody old hag, begone!" Here the bands of slumber were broken by the intensity of his feelings, and with a wild expression of countenance and a frame shaking with emotion, he started from the bench, and stood trembling before me.

Though convinced that he was a criminal, I could not help pitying him from the forlorn appearance he now exhibited. As soon as he had collected his wandering ideas, it seemed as if he read in my countenance, the mingled sentiments of pity and abhorrence, with which I regarded him. Looking anxiously around, and seeing that we were alone, he drew the corner of the bench towards me, and sitting down, with an apparent effort to command his feelings, thus addressed me. His tone of voice was calm, and distinct; and his countenance, though deadly pale, was composed.

"I see, Sir, that from what I am conscious of having uttered in my disturbed sleep, you suspect me of some horrid crime. You are right. My conscience convicts me, and an awful nightly visitation, worse than the waking pangs of remorse, compels me to confess it. Yes, I am a murderer. I have been the unhappy cause of blotting out the life of a fellow being from the page of human existence. In these pallid features, you may read enstamped, in the same characters which the first murderer bore upon his brow, Guilt—guilt—guilt!"

Here the poor young man paused, evidently agitated by strong internal emotion. Collecting himself, however, in a few moments, he thus continued.

"Yet still, when you have heard my sad story, I think you will bestow upon me your pity. I feel that there is no peace for me, until I have disburthened my mind. Your countenance promises sympathy. Will you listen to my unhappy narrative?"

My curiosity being strongly excited by this strange exordium, I told him I was ready to hear whatever he had to communicate. Upon this, he proceeded as follows:—

"My name is Hippocrates Jenkins. I was born in Nantucket, but my father emigrated to these parts when I was young. I grew up in one of the most flourishing villages on the borders of the canal. My father and mother both dying of the lake fever, I was bound apprentice to an eminent operative in the boot and shoe making line, who had lately come from New York. Would that I had remained content with this simple and useful profession. Would that I had stuck to my waxed ends and awl, and never undertaken to cobble up people's bodies. But my legs grew tired of being trussed beneath my haunches; my elbows wearied with their monotonous motion; my eyes became dim with gazing forever upon the dull brick wall which faced our shop window; and my whole heart was sick of my sedentary, and, as I foolishly deemed it, particularly mean occupation. My time was nearly expired, and I had long resolved, should any opportunity offer of getting into any other employment, I would speedily embrace it.

"I had always entertained a predilection for the study of medicine. What had given my mind this bias, I know not. Perhaps it was the perusal of an old volume of Doctor Buchan,[3] over whose pages it was the delight of my youthful fancy to pore. Perhaps it was the oddness of my Christian cognomen, which surely was given me by my parents in a prophetic hour. Be this as it may, the summit of my earthly happiness was to be a doctor. Conceive then my delighted surprise, one Saturday evening, after having carried home a pair of new white-topped boots for Doctor Ephraim Ramshorne, who made the care of bodies his care, in the village, to hear him ask me, how I should like to be a doctor. He then very generously offered to take me as a student. From my earliest recollections, the person and character of Doctor Ramshorne had been regarded by me with the most profound and awful admiration. Time out of mind the successful practitioner for many miles around, I had looked upon him as the *beau idéal* of a doctor—a very Apollo in the healing art. When I speak of him, however, as the *successful* practitioner, I mean it not to be inferred that death was less busy in his doings, or funerals scarcer during his dynasty; but only that he had, by some means or other, contrived to force all those who had ventured to contest the palm with him, to quit the field. He was large and robust in person, and his ruby visage showed that if he grew fat upon drugs, it was not by swallowing them himself. It was never exactly ascertained from what college the Doctor had received his diploma; nor was he very forward to exhibit his credentials. When hard pressed, however, he would produce a musty old roll of parchment, with a red seal as broad as the palm of his hand, which looked as if it might have been the identical diploma of the great Boerhaave himself, and some cramped manuscript of a dozen pages, in an unknown tongue, said by the Doctor to be his Greek thesis. These documents were enough to satisfy the doubts of the most sceptical.[4] By the simple country people, far and near, the Doctor was regarded, in point of occult knowledge and skill, as a second Faustus.[5] It is true the village lawyer, a rival in popularity, used to whisper, that the Doctor's Greek thesis was nothing but a bundle of prescriptions for the bots, wind-galls, spavins, and other veterinary complaints, written in high Dutch by a Hessian horse doctor; that the diploma was all a sham, and that Ephraim was no more a doctor than his jackass. But these assertions were all put down to the score of envy on the part of the lawyer. Be this as it may, on the strength of one or two remarkable cures, which he was said to have performed, and by dint of wheedling some and bullying others, it was certain that Ramshorne had worked himself into very good practice. The Doctor united in his own person, the attributes of apothecary and physician; and as he vended, as well as prescribed his own drugs, it was not his interest to stint his patients in their enormous boluses, or nauseous draughts. His former medical student had been worried into a consumption over the mortar and pestle; in consequence of which he had pitched upon me for his successor.

"By the kindness of a few friends I was fitted out with the necessary requisitions for my metamorphosis. The Doctor required no fee, and, in consideration of certain little services to be rendered him, such as taking care of his horse, cleaning his boots, running errands, and doing little jobs about the house, had promised to board and lodge me, besides giving me my professional education. So with a rusty suit of black, and an old plaid cloak, behold equipped the disciple of Esculapius.

"I cannot describe my elation of mind, when I found myself fairly installed in the Doctor's office. Golden visions floated before my eyes. I fancied my fortune already made, and blessed my happy star, that had fallen under the benign influence of so munificent a patron.

"The Doctor's office, as it was called *par excellence,* was a little nook of a room, communicating with a larger apartment denominated the shop. The paraphernalia of this latter place had gotten somewhat into disorder since the last student had gone away, and I soon learnt that it was to be my task to arrange the heterogeneous mass of bottles, boxes, and gallipots, that were strewed about in promiscuous confusion. In the office, there was a greater appearance of order. A small regiment of musty looking books, were drawn up in line upon a couple of shelves, where, to judge from the superincumbent strata of dust, they appeared to have peacefully reposed for many years. A rickety wooden clock, which the

Doctor had taken in part payment from a peddler, and the vital functions of which, to use his own expression, had long since ceased to act, stood in one corner. A mouldy plaster bust of some unknown worthy, a few bottles of pickled, and one or two dried specimens of morbid anatomy, a small chest of drawers, a table, and a couple of chairs, completed the furniture of this *sanctum*. The single window commanded a view of the churchyard, in which, it was said, many of the Doctor's former patients were quietly slumbering. With a feeling of reverence I ventured to dislodge one of the dusty tomes, and began to try to puzzle out the hard words with which it abounded; when suddenly, as if he had been conjured back, like the evil one by Cornelius Agrippa's book, the Doctor made his appearance. With a gruff air, he snatched the volume from my hands, and telling me not to meddle with what I could not understand, bade me go and take care of his horse, and make haste back, as he wanted me to spread a pitch plaster, and carry the same, with a bottle of his patent catholicon, to farmer Van Pelt, who had the rheumatism. On my return, I was ordered by Mrs. Ramshorne to split some wood, and kindle a fire in the parlor, as she expected company; after which Miss Euphemia Ramshorne, a sentimental young lady, who was as crooked in person and crabbed in temper as her own name, despatched me to the village circulating library, in quest of the Mysteries of Udolpho. I soon found out that my place was no sinecure. The greater part of my time was occupied in compounding certain quack medicines of Ramshorne's own invention, from which he derived great celebrity, and no inconsiderable profit. Besides his patent catholicon, and universal panacea, there was his anti-pertusso-balsamico drops, his patent calorific refrigerating anodyne, and his golden restorative of nature. Into the business of compounding these, and other articles with similar high-sounding titles, I was gradually initiated, and soon acquired so much skill in their manipulation, that my services became indispensable to my master; so much so, that he was obliged to hire a little negro to take care of his horse, and clean his boots. What chiefly reconciled me to the drudgery of the shop, was the seeing how well the Doctor got paid for his villainous compounds. A mixture of a little brick dust, rosin, and treacle, dignified with the title of the anthelminthic amalgam, he sold for half a dollar; and a bottle of vinegar and alum, with a little rose water to give it a flavor, yclept the anti-scrofulous abstergent lotion, brought twice that sum. I longed for the day when I should dispense my own medicines, and in my hours of castle-building, looked forward to fortunes far beyond those of the renowned Dr. Solomon. Alas! my fond hopes have been blighted in their bud. I have drunk deeply of the nauseous draught of adversity, and been forced to swallow many bitter pills of disappointment. But I find I am beginning to smell of the shop. I must return to my sad tale. The same accident, which not unfrequently before had put a sudden stop to the Doctor's patients taking any more of his nostrums, at length prevented him from reaping any longer their golden harvest. One afternoon, after having dined with his friend, Squire Gobbledown, he came home, and complained of not feeling very well. By his directions, I prepared for him some of his elixir sanitatis, composed of brandy and bitters, of which he took an inordinate dose. Shortly after, he was seized with a fit of apoplexy, and before bedtime, in spite of all the drugs in the shop, which I poured down with unsparing hand, he had breathed his last. In three days, Ramshorne was quietly deposited in the churchyard, in the midst of those he had sent there before him.

"Having resided with the Doctor for several years, I had become pretty well known throughout the neighborhood, particularly among the old ladies, whose good graces I had always sedulously cultivated. I accordingly resolved to commence quacking—I mean practicing—on my own account. Having obtained my late master's stock of drugs from his widow at an easy rate, and displaying my own name in golden letters as his successor, to work I went, with the internal resolve that where Ramshorne had given one dose, I would give six.

"For a time, Fortune seemed to smile upon me, and everything went on well. All the old women were loud in sounding my praises, far and near. The medicaments of my master, continued to be in demand, and treacle, brick dust, and alum came to a good market. Some drawbacks, however, I occasionally met with. Having purchased the patent right of one of Thompson's steam baths, in my first experiment I came near flaying alive a rheumatic tanner, who had submitted himself to the operation. By an unfortunate mistake in regulating the steam, he was nearly parboiled; and it was supposed that the thickness of his hide alone preserved his vitals uninjured. I was myself threatened with the fate of Marsyas, by the enraged sufferer; which he was happily prevented from attempting to inflict, by a return of his malady, which has never since left him. I however after this gave up steaming, and confined myself to regular practice. At length, either the charm of novelty wearing off, or people beginning to discover the inefficacy of the old nostrums, I was obliged to exert my wit to invent new ones. These I generally took the precaution to try upon cats or dogs, before using them upon the human system. They were, however, mostly of an innocent nature, and I satisfied my conscience with the reflection, that if they did no good, they could at least do no harm. Happy would it have been for me, could I always have done thus. Meeting with success in my first efforts, I by degrees ventured upon more active ingredients. At length in an evil hour,

I invented a curious mixture, composed of forty-nine different articles. This I dubbed in high-flowing terms 'The Antidote to Death, or the Eternal Elixir of Longevity;' knowing full well, that though

'A rose might smell as sweet by any other name,'

yet would not my drugs find as good a sale under a more humble title. This cursed compound proved the antidote to all my hopes of success. Besides forcing me to quit the village in a confounded hurry, it has embittered my life ever since, and reduced me to the ragged and miserable plight in which you see me.

"I dare say you have met with that species of old women, so frequent in all country towns, who, seeming to have outlived the common enjoyments of life, and outworn the ordinary sources of excitement, seek fresh stimulus in scenes of distress, and appear to take a morbid pleasure in beholding the varieties of human suffering and misery. One of the most noted characters in the village was an old beldame of this description. Granny Gordon, so she was familiarly denominated, was the rib of the village Vulcan, and the din of her eternal tongue was only equalled by the ringing of her husband's anvil. Thin and withered away in person and redolent with snuff, she bore no small resemblance to a newly-exhumed mummy, and to all appearance promised to last as long as one of those ancient dames of Egypt. Not a death, a burial, a fit of sickness, a casualty, nor any of the common calamities of life ever occurred in the vicinity, but Granny Gordon made it her especial business to be present. Wrapped in an old scarlet cloak,—that hideous cloak! the thought of it makes me shudder—she might be seen hovering about the dwelling of the sick. Watching her opportunity, she would make her way into the patient's chamber, and disturb his repose with long dismal stories and ill-boding predictions; and if turned from the house, which was not unfrequently the case, she would depart, muttering threats and abuse.

"As the Indians propitiate the favor of the devil, so had I, in my eagerness to acquire popularity, made a firm friend and ally, though rather a troublesome one, of this old woman. She was one of my best customers, and, provided it was something new, and had a high-sounding name to recommend it, would take my most nauseous compounds with the greatest relish. Indeed the more disgusting was the dose, the greater in her opinion was its virtue.

"I had just corked the last bottle of my antidote, when a message came to tell me that Granny Gordon had one of her old fits, and wanted some new doctor-stuff as the old physic didn't do her any more good. Not having yet given my new pharmaceutic preparation a trial, I felt a little doubtful about its effects, but trusting to the toughness of the old woman's system, I ventured to send a potion, with directions to take it cautiously. Not many minutes had elapsed, before the messenger returned, in breathless haste, to say that Mrs. Gordon was much worse, and that though she had taken all the stuff, they believed she was dying. With a vague foreboding of evil, I seized my hat, and hastened to the blacksmith's. On entering the chamber my eyes were greeted with a sad spectacle. Granny Gordon, bolstered up in the bed, holding in her hand the bottle I had sent her, drained of its contents, sat gasping for breath, and occasionally agitated by strong convulsions. A cold sweat rested on her forehead, her eyes seemed dim and glazed, her nose, which was usually of a ruby hue, was purple and peaked, and her whole appearance evidently betokened approaching dissolution.

"Around the bed were collected some half dozen withered beldames, who scowled upon me, as I entered, with ill-omened visages. Her husband, a drunken brute, who used to beat his better half six times a week, immediately began to load me with abuse, accusing me of having poisoned his dear, dear wife, and threatening to be the death of me, if she died.

"My conscience smote me. I felt stupefied and bewildered, and knew not which way to turn. At this moment, the patient perceiving me, with a hideous contortion of countenance, the expression of which I shall carry to my dying hour, and a voice between a scream and a groan, held up the empty bottle, and exclaimed, 'This is your doing, you villainous quack you' (here she was seized with hiccup);—'you have poisoned me, you have' (here fearful spasms shook her whole frame);—'but I'll be revenged; day and night my ghost shall haunt'—here her voice became inarticulate, and shaking her withered arm at me, she fell back, and, to my extreme horror, gave up the ghost. This was too much for my nerves. I rushed from the house, and ran home with the dying curse ringing in my ears, fancying that I saw her hideous physiognomy, grinning from every bush and tree that I passed. Knowing that as soon as the noise of this affair should get abroad, the village would be too hot to hold me, I resolved to decamp as silently as possible. First throwing all my recently manufactured anodyne into the canal, that it should not rise in judgment against me, I made up a little bundle of clothes, and taking my seat in the mail stage, which was passing at the time and fortunately empty, in a couple of days I found myself in the great city of New York. Having a little money with me, I hired a mean apartment in an obscure part of the city, in the hope that I might remain concealed till all search after me should be over, when I might find some opportunity of getting employment, or of resuming my old profession, under happier auspices. By degrees the few dollars I brought with me were expended, and after pawning my watch and some of my clothes, I found myself reduced to the last shilling.

But not the fear of impending starvation, nor the dread of a jail, are to be compared to the horrors I nightly suffer. Granny Gordon has been as good as her word. Every night, at the solemn hour of twelve" (here he looked fearfully around)—"her ghost appears to me, wrapped in a red cloak, with her gray hairs streaming from beneath an old nightcap of the same color, brandishing the vial, and accusing me of having poisoned her. These visitations have at length become so insupportable, that I have resolved to return and give myself up to justice; for I feel that hanging itself is better than this state of torment."

Here the young man ceased. I plainly saw that he was a little disordered in his intellect. To comfort him, however, I told him, that if he had killed fifty old women, they could do nothing to him, if he had done it professionally. And as for the ghost, we would take means to have that put at rest, when we reached Utica.

About the gray of the morning, we arrived at the place of our destination. My *protégé,* having unburthened his mind, seemed more at his ease, and taking a mint julep, prepared to accompany me on shore. As we were leaving the boat, several persons in a wagon drove down to the wharf. As soon as my companion observed them, he exclaimed with a start of surprise, "Hang me, if there is n't old Graham the sheriff, with lawyer Dickson, and Bill Gordon come to take me." As he spoke, his foot slipping, he lost his balance, and fell backwards into the canal. We drew him from the water, and as soon as the persons in the wagon perceived him, they one and all sprang out, and ran up with the greatest expressions of joyful surprise. "Why Hippy, my lad," exclaimed the sheriff, "where have you been? All our town has been in a snarl about you. We all supposed you had been forcibly abducted. Judge Bates offered a reward of twenty dollars for your corpse. We have dragged the canal for more than a mile, and found a mass of bottles, which made us think you had been spirited away. Betsey Wilkins made her affidavit that she heard Bill Gordon swear that he would take your life, and here you see we have brought him down to have his trial. But come, come, jump in the wagon, we'll take you up to the tavern, to get your duds dried, and tell you all about it."

Here a brawny fellow with a smutty face, who I found was Gordon the black-smith, came up, and shaking Hippocrates by the hand, said, "By goles, Doctor, I am glad to see you. If you had n't come back, I believe it would have gone hard with me. Come, man, you must forgive the hard words I gave you. My old woman soon got well of her fit, after you went away, and says she thinks the stuff did her a mortal sight o' good."

It is impossible to describe the singular expression the countenance of the young man now exhibited. For some time he stood in mute amazement, shaking with cold, and gazing alternately at each of his friends as they addressed him; and it required their reiterated assurance to convince him, that Granny Gordon was still in the land of the living, and that he had not been haunted by a veritable ghost.

Wishing to obtain a further explanation of this strange scene, I accompanied them to the tavern. A plain-looking man in a farmer's dress, who was of the party, confirmed what the blacksmith had said, as to the supposed death of his wife, and her subsequent recovery. "She was only in a swoond," said he, "but came to, soon after the Doctor had left her." He added that it was his private opinion that she would now last forever. He spoke of Hippocrates as a "'nation smart doctor, who had a power of larning, but gave severe doses."

After discussing a good breakfast, my young friend thanked me for the sympathy and interest I had taken in his behalf. He told me he intended returning to the practice of his profession. I admonished him to be more careful in the exhibition of his patent medicines, telling him that all old women had not nine lives. He shook hands with me, and, gayly jumping into the wagon, rode off with his friends.

1. How would you describe Hippocrates Jenkins's medical training—and Doctor Ramshorne, who trained him? How does Hippocrates qualify his comment that Doctor Ramshorne was a "successful doctor," and what does this say about Hippocrates's motivations to become a physician? Make a list of the various reasons one might pursue a career in health care or biomedical research today. Do one's motivations influence the ability to practice ethically and effectively? Provide an example to support your claim.

2. Hippocrates Jenkins tells himself that even if his medicines "did no good, they could at least do no harm." Does this reflect the true spirit behind the Hippocratic Oath? Does it reflect cultural realities of the time? Which cultural and professional realities of the twenty-first century are most challenging to young medical professionals who seek to uphold this oath? Provide specific examples.

3. This tale touches on issues of medical error disclosure. Should health care providers disclose all errors, even those that do not result in adverse events? Discuss the role empathetic communication can play in preventing errors and providing appropriate care to patients and family members when an error does occur.

4. "I feel that there is no peace for me, until I have disburthened my mind," Hippocrates Jenkins tells the narrator. "Your countenance promises sympathy. Will you listen to my unhappy narrative?" His words imply that he sees a connection between storytelling and healing. Do you believe storytelling in the form of reflective or creative writing or other art therapies can play a role in healing? Research what evidence exists to support arts–in–health care therapies.

5. Compare Hippocrates Jenkins's motivations to those of Dr. Rappaccini of "Rappaccini's Daughter," or Aylmer of "The Birthmark." Also, compare the female victims in those stories to Granny Gordon. Give special consideration to the ways each character's age shapes our response to her situation.

The Minister's Black Veil
A Parable*
1836, *Token*

"'The Minister's Black Veil' is a masterly composition," Edgar Allan Poe wrote in his 1842 review of Hawthorne's Twice-Told Tales, "of which the sole defect is that to the rabble its exquisite skill will be caviare. The obvious *meaning of this article will be found to smother its insinuated one. The* moral *put into the mouth of the dying minister will be supposed to convey the* true *import of the narrative; and that a crime of dark dye (having reference to the 'young lady'), has been committed, is a point which only minds congenial with that of the author will perceive" (Poe 1984, 574–75). Poe, who shared Hawthorne's interest in analyzing pathological behavior, suggests that while most readers would buy the explanation the minister provides at the story's end, more "congenial" minds would find another, more complicated root of illness.*

A century and a half after Poe's bold assertion, literary scholar Carl Ostrowski (1998) took up Poe's challenge to read better than the "rabble," proposing that the minister's veil is not merely symbolic but that it hides the signs of the sexually contracted disease of syphilis, inviting readers to investigate the sociohistorical context of the tale. The idea that Hooper might suffer from a physical rather than a moral illness is intriguing for a few reasons. First, it reminds us that Hawthorne was "anxiously positioned" (Goldman 2004, 30) between two approaches to illness in the nineteenth century—first, the religious view that proposed physical illness originated in spiritual dysfunction and second, the evolving view of medicine that suggested all mental illness was derived from a physiological source. Readers might be interested in pursuing dual readings of the tale, using each of these nineteenth-century medical models to diagnose the cause of the minister's suffering.[1]

And even though we like to think science and medicine have come a long

---

*Another clergyman in New England, Mr. Joseph Moody, of York, Maine, who died about eighty years since, made himself remarkable by the same eccentricity that is here related of the Reverend Mr. Hooper. In his case, however, the symbol had a different import. In early life he had accidentally killed a beloved friend; and from that day till the hour of his own death, he hid his face from men.

way since the Puritans, readers might also consider whether we perpetuate the idea that certain physical illnesses are a sign of moral failing. Are some diseases considered shameful or immoral in the United States today? If so, what are they and what types of marginalization or stigmatization do patients with those diseases experience? How do these stigmas influence health—and what might we do in the twenty-first century to address them?

•

THE sexton stood in the porch of Milford meeting-house, pulling busily at the bell-rope. The old people of the village came stooping along the street. Children, with bright faces, tripped merrily beside their parents, or mimicked a graver gait, in the conscious dignity of their Sunday clothes. Spruce bachelors looked sidelong at the pretty maidens, and fancied that the Sabbath sunshine made them prettier than on week days. When the throng had mostly streamed into the porch, the sexton began to toll the bell, keeping his eye on the Reverend Mr. Hooper's door. The first glimpse of the clergyman's figure was the signal for the bell to cease its summons.

"But what has good Parson Hooper got upon his face?" cried the sexton in astonishment.

All within hearing immediately turned about, and beheld the semblance of Mr. Hooper, pacing slowly his meditative way towards the meeting-house. With one accord they started, expressing more wonder than if some strange minister were coming to dust the cushions of Mr. Hooper's pulpit.

"Are you sure it is our parson?" inquired Goodman Gray of the sexton.

"Of a certainty it is good Mr. Hooper," replied the sexton. "He was to have exchanged pulpits with Parson Shute, of Westbury; but Parson Shute sent to excuse himself yesterday, being to preach a funeral sermon."

The cause of so much amazement may appear sufficiently slight. Mr. Hooper, a gentlemanly person, of about thirty, though still a bachelor, was dressed with due clerical neatness, as if a careful wife had starched his band, and brushed the weekly dust from his Sunday's garb. There was but one thing remarkable in his appearance. Swathed about his forehead, and hanging down over his face, so low as to be shaken by his breath, Mr. Hooper had on a black veil. On a nearer view it seemed to consist of two folds of crape, which entirely concealed his features, except the mouth and chin, but probably did not intercept his sight, further than to give a darkened aspect to all living and inanimate things. With this gloomy shade before him, good Mr. Hooper walked onward, at a slow and quiet pace, stooping somewhat, and looking on the ground, as is customary with abstracted men, yet nodding kindly to those of his parishioners who still waited on the meeting-house steps. But so wonder-struck were they that his greeting hardly met with a return.

"I can't really feel as if good Mr. Hooper's face was behind that piece of crape," said the sexton.

"I don't like it," muttered an old woman, as she hobbled into the meeting-house. "He has changed himself into something awful, only by hiding his face."

"Our parson has gone mad!" cried Goodman Gray, following him across the threshold.

A rumor of some unaccountable phenomenon had preceded Mr. Hooper into the meeting-house, and set all the congregation astir. Few could refrain from twisting their heads towards the door; many stood upright, and turned directly about; while several little boys clambered upon the seats, and came down again with a terrible racket. There was a general bustle, a rustling of the women's gowns and shuffling of the men's feet, greatly at variance with that hushed repose which should attend the entrance of the minister. But Mr. Hooper appeared not to notice the perturbation of his people. He entered with an almost noiseless step, bent his head mildly to the pews on each side, and bowed as he passed his oldest parishioner, a white-haired great-grandsire, who occupied an arm-chair in the centre of the aisle. It was strange to observe how slowly this venerable man became conscious of something singular in the appearance of his pastor. He seemed not fully to partake of the prevailing wonder, till Mr. Hooper had ascended the stairs, and showed himself in the pulpit, face to face with his congregation, except for the black veil. That mysterious emblem was never once withdrawn. It shook with his measured breath, as he gave out the psalm; it threw its obscurity between him and the holy page, as he read the Scriptures; and while he prayed, the veil lay heavily on his uplifted countenance. Did he seek to hide it from the dread Being whom he was addressing?

Such was the effect of this simple piece of crape, that more than one woman of delicate nerves was forced to leave the meeting-house. Yet perhaps the pale-faced congregation was almost as fearful a sight to the minister, as his black veil to them.

Mr. Hooper had the reputation of a good preacher, but not an energetic one: he strove to win his people heavenward by mild, persuasive influences, rather than to drive them thither by the thunders of the Word. The sermon which he now delivered was marked by the same characteristics of style and manner as the general series of his pulpit oratory. But there was something, either in the sentiment of the discourse itself, or in the imagination of the auditors, which made it greatly the most powerful effort that they had ever heard from their pastor's lips. It was tinged, rather more darkly than usual, with the gentle gloom of Mr. Hooper's temperament. The subject had reference to secret sin, and those sad mysteries which we hide from our nearest and dearest, and would fain conceal from our own consciousness, even forgetting that the Omniscient can detect them. A subtle power was breathed into his words. Each member of the congregation, the most innocent girl, and the man of hardened breast, felt as if the preacher had crept upon them, behind his awful veil, and discovered their hoarded iniquity of deed or thought. Many spread their clasped hands on their bosoms. There was nothing terrible in what Mr. Hooper said, at least, no violence; and yet, with every tremor of his melancholy voice, the hearers quaked. An unsought pathos came hand in hand with awe. So sensible were the audience of some unwonted attribute in their minister, that they longed for a breath of wind to blow aside the veil, almost believing that a stranger's visage would be discovered, though the form, gesture, and voice were those of Mr. Hooper.

At the close of the services, the people hurried out with indecorous confusion, eager to communicate their pent-up amazement, and conscious of lighter spirits the moment they lost sight of the black veil. Some gathered in little circles, huddled closely together, with their mouths all whispering in the centre; some went homeward alone, wrapt in silent meditation; some talked loudly, and profaned the Sabbath day with ostentatious laughter. A few shook their sagacious heads, intimating that they could penetrate the mystery; while one or two affirmed that there was no mystery at all, but only that Mr. Hooper's eyes were so weakened by the midnight lamp, as to require a shade. After a brief interval, forth came good Mr. Hooper also, in the rear of his flock. Turning his veiled face from one group to another, he paid due reverence to the hoary heads, saluted the middle aged with kind dignity as their friend and spiritual guide, greeted the young with mingled authority and love, and laid his hands on the little children's heads to bless them. Such was always his custom on the Sabbath day. Strange and bewildered looks repaid him for his courtesy. None, as on former occasions, aspired to the honor of walking by their pastor's side. Old Squire Saunders, doubtless by an accidental lapse of memory, neglected to invite Mr. Hooper to his table, where the good clergyman had been wont to bless the food, almost every Sunday since his settlement. He returned, therefore, to the parsonage, and, at the moment of closing the door, was observed to look back upon the people, all of whom had their eyes fixed upon the minister. A sad smile gleamed faintly from beneath the black veil, and flickered about his mouth, glimmering as he disappeared.

"How strange," said a lady, "that a simple black veil, such as any woman might wear on her bonnet, should become such a terrible thing on Mr. Hooper's face!"

"Something must surely be amiss with Mr. Hooper's intellects," observed her husband, the physician of the village. "But the strangest part of the affair is the effect of this vagary, even on a sober-minded man like myself. The black veil, though it covers only our pastor's face, throws its influence over his whole person, and makes him ghostlike from head to foot. Do you not feel it so?"

"Truly do I," replied the lady; "and I would not be alone with him for the world. I wonder he is not afraid to be alone with himself!"

"Men sometimes are so," said her husband.

The afternoon service was attended with similar circumstances. At its conclusion, the bell tolled for the funeral of a young lady. The relatives and friends were assembled in the house, and the more distant acquaintances stood about the door, speaking of the good qualities of the deceased, when their talk was interrupted by the appearance of Mr. Hooper, still covered with his black veil. It was now an appropriate emblem. The clergyman stepped into the room where the corpse was laid, and bent over the coffin, to take a last farewell of his deceased parishioner. As he stooped, the veil hung straight down from his forehead, so that, if her eyelids had not been closed forever, the dead maiden might have seen his face. Could Mr. Hooper be fearful of her glance, that he so hastily caught back the black veil? A person who watched the interview between the dead and living, scrupled not to affirm, that, at the instant when the clergyman's features were disclosed, the corpse had slightly shuddered, rustling the shroud and muslin cap, though the countenance retained the composure of death. A superstitious old woman was the only witness of this prodigy. From the coffin Mr. Hooper passed into the chamber of the mourners, and thence to the head of the staircase, to make the funeral prayer. It was a tender and heart-dissolving prayer, full of sorrow, yet so imbued with celestial hopes, that the music of a heavenly harp, swept by the fingers of the dead, seemed faintly to be heard among the saddest accents of the minister. The people trembled, though they but darkly understood him when he prayed that they, and himself, and all of mortal race, might be ready, as he trusted this young maiden had been, for the dreadful hour that should snatch the veil from their faces. The bearers went heavily forth, and the mourners followed, saddening all the street, with the dead before them, and Mr. Hooper in his black veil behind.

"Why do you look back?" said one in the procession to his partner.

"I had a fancy," replied she, "that the minister and the maiden's spirit were walking hand in hand."

"And so had I, at the same moment," said the other.

That night, the handsomest couple in Milford village were to be joined in wedlock. Though reckoned a melancholy man, Mr. Hooper had a placid cheerfulness for such occasions, which often excited a sympathetic smile where livelier merriment would have been thrown away. There was no quality of his disposition which made him more beloved than this. The company at the wedding awaited his arrival with impatience, trusting that the strange awe, which had gathered over him throughout the day, would now be dispelled. But such was not the result. When Mr. Hooper came, the first thing that their eyes rested on was the same horrible black veil, which had added deeper gloom to the funeral, and could portend nothing but evil to the wedding. Such was its immediate effect on the guests that a cloud seemed to have rolled duskily from beneath the black crape, and dimmed the light of the candles. The bridal pair stood up before the minister. But the bride's cold fingers quivered in the tremulous hand of the bridegroom, and her deathlike paleness caused a whisper that the maiden who had been buried a few hours before was come from her grave to be married. If ever another wedding were so dismal, it was that famous one where they tolled the wedding knell. After performing the ceremony, Mr. Hooper raised a glass of wine to his lips, wishing happiness to the new-married couple in a strain of mild pleasantry that ought to have brightened the features of the guests, like a cheerful gleam from the hearth. At that instant, catching a glimpse of his figure in the looking-glass, the black veil involved his own spirit in the horror with which it overwhelmed all others. His frame shuddered, his lips grew white, he spilt the untasted wine upon the carpet, and rushed forth into the darkness. For the Earth, too, had on her Black Veil.

The next day, the whole village of Milford talked of little else than Parson Hooper's black veil. That, and the mystery concealed behind it, supplied a topic for discussion between acquaintances meeting in the street, and good women gossiping at their open windows. It was the first item of news that the tavern-keeper told to his guests. The children babbled of it on their way to school. One imitative little imp covered his face with an old black handker-chief, thereby so affrighting his playmates that the panic seized himself, and he well-nigh lost his wits by his own waggery.

It was remarkable that of all the busybodies and impertinent people in the parish, not one ventured to put the plain question to Mr. Hooper, wherefore he did this thing. Hitherto, whenever there appeared the slightest call for such interference, he had never lacked advisers, nor shown himself adverse to be guided by their judgment. If he erred at all, it was by so painful a degree of self-distrust, that even the mildest censure would lead him to consider an indifferent action as a crime. Yet, though so well acquainted with this amiable weakness, no individual among his parishioners chose to make the black veil a subject of friendly remonstrance. There was a feeling of dread, neither plainly confessed nor carefully concealed, which caused each to shift the responsibility upon another, till at length it was found expedient to send a deputation of the church, in order to deal with Mr. Hooper about the mystery, before it should grow into a scandal. Never did an embassy so ill discharge its duties. The minister received them with friendly courtesy, but became silent, after they were seated, leaving to his visitors the whole burden of introducing their important business. The topic, it might be supposed, was obvious enough. There was the black veil swathed round Mr. Hooper's forehead, and concealing every feature above his placid mouth, on which, at times, they could perceive the glimmering of a melancholy smile. But that piece of crape, to their imagination, seemed to hang down before his heart, the symbol of a fearful secret between him and them. Were the veil but cast aside, they might speak freely of it, but not till then. Thus they sat a considerable time, speechless, confused, and shrinking uneasily from Mr. Hooper's eye, which they felt to be fixed upon them with an invisible glance. Finally, the deputies returned abashed to their constituents, pronouncing the matter too weighty to be handled, except by a council of the churches, if, indeed, it might not require a general synod.

But there was one person in the village unappalled by the awe with which the black veil had impressed all beside herself. When the deputies returned without an explanation, or even venturing to demand one, she, with the calm energy of her character, determined to chase away the strange cloud that appeared to be settling round Mr. Hooper, every moment more darkly than before. As his plighted wife, it should be her privilege to know what the black veil concealed. At the minister's first visit, therefore, she entered upon the subject with a direct simplicity, which made the task easier both for him and her. After he had seated himself, she fixed her eyes steadfastly upon the veil, but could discern nothing of the dreadful gloom that had so overawed the multitude: it was but a double fold of crape, hanging down from his forehead to his mouth, and slightly stirring with his breath.

"No," said she aloud, and smiling, "there is nothing terrible in this piece of crape, except that it hides a face which I am always glad to look upon. Come, good sir, let the sun shine from behind the cloud. First lay aside your black veil: then tell me why you put it on."

Mr. Hooper's smile glimmered faintly.

"There is an hour to come," said he, "when all of us shall cast aside our veils. Take it not amiss, beloved friend, if I wear this piece of crape till then."

"Your words are a mystery, too," returned the young lady. "Take away the veil from them, at least."

"Elizabeth, I will," said he, "so far as my vow may suffer me. Know, then, this veil is a type and a symbol, and I am bound to wear it ever, both in light and darkness, in solitude and before the gaze of multitudes, and as with strangers, so with my familiar friends. No mortal eye will see it withdrawn. This dismal shade must separate me from the world: even you, Elizabeth, can never come behind it!"

"What grievous affliction hath befallen you," she earnestly inquired, "that you should thus darken your eyes forever?"

"If it be a sign of mourning," replied Mr. Hooper, "I, perhaps, like most other mortals, have sorrows dark enough to be typified by a black veil."

"But what if the world will not believe that it is the type of an innocent sorrow?" urged Elizabeth. "Beloved and respected as you are, there may be whispers that you hide your face under the consciousness of secret sin. For the sake of your holy office, do away this scandal!"

The color rose into her cheeks as she intimated the nature of the rumors that were already abroad in the village. But Mr. Hooper's mildness did not forsake him. He even smiled again—that same sad smile, which always appeared like a faint glimmering of light, proceeding from the obscurity beneath the veil.

"If I hide my face for sorrow, there is cause enough," he merely replied; "and if I cover it for secret sin, what mortal might not do the same?"

And with this gentle, but unconquerable obstinacy did he resist all her entreaties. At length Elizabeth sat silent. For a few moments she appeared lost in thought, considering, probably, what new methods might be tried to withdraw her lover from so dark a fantasy, which, if it had no other meaning, was perhaps a symptom of mental disease. Though of a firmer character than his own, the tears rolled down her cheeks. But, in an instant, as it were, a new feeling took the place of sorrow: her eyes were fixed insensibly on the black veil, when, like a sudden twilight in the air, its terrors fell around her. She arose, and stood trembling before him.

"And do you feel it then, at last?" said he mournfully.

She made no reply, but covered her eyes with her hand, and turned to leave the room. He rushed forward and caught her arm.

"Have patience with me, Elizabeth!" cried he, passionately. "Do not desert me, though this veil must be between us here on earth. Be mine, and hereafter there shall be no veil over my face, no darkness between our souls! It is but a mortal veil—it is not for eternity! O! you know not how lonely I am, and how frightened, to be alone behind my black veil. Do not leave me in this miserable obscurity forever!"

"Lift the veil but once, and look me in the face," said she.

"Never! It cannot be!" replied Mr. Hooper.

"Then farewell!" said Elizabeth.

She withdrew her arm from his grasp, and slowly departed, pausing at the door, to give one long shuddering gaze, that seemed almost to penetrate the mystery of the black veil. But, even amid his grief, Mr. Hooper smiled to think that only a material emblem had separated him from happiness, though the horrors, which it shadowed forth, must be drawn darkly between the fondest of lovers.

From that time no attempts were made to remove Mr. Hooper's black veil, or, by a direct appeal, to discover the secret which it was supposed to hide. By persons who claimed a superiority to popular prejudice, it was reckoned merely an eccentric whim, such as often mingles with the sober actions of men otherwise rational, and tinges them all with its own semblance of insanity. But with the multitude, good Mr. Hooper was irreparably a bugbear. He could not walk the street with any peace of mind, so conscious was he that the gentle and timid would turn aside to avoid him, and that others would make it a point of hardihood to throw themselves in his way. The impertinence of the latter class compelled him to give up his customary walk at sunset to the burial ground; for when he leaned pensively over the gate, there would always be faces behind the gravestones, peeping at his black veil. A fable went the rounds that the stare of the dead people drove him thence. It grieved him, to the very depth of his kind heart, to observe how the children fled from his approach, breaking up their merriest sports, while his melancholy figure was yet afar off. Their instinctive dread caused him to feel more strongly than aught else, that a preternatural horror was interwoven with the threads of the black crape. In truth, his own antipathy to the veil was known to be so great, that he never willingly passed before a mirror, nor stooped to drink at a still fountain, lest, in its peaceful bosom, he should be affrighted by himself. This was what gave plausibility to the whispers, that Mr. Hooper's conscience tortured him for some great crime too horrible to be entirely concealed, or otherwise than so obscurely intimated. Thus, from beneath the black veil, there rolled a cloud into the sunshine, an ambiguity of sin or sorrow, which enveloped the poor minister, so that love or sympathy could never reach him. It was said that ghost and fiend consorted with him there. With self-shudderings and outward terrors, he walked continually in its shadow, groping darkly within his own soul, or gazing through a medium that saddened the whole world. Even the lawless wind, it was believed, respected his dreadful secret, and never blew aside the veil. But still good Mr. Hooper sadly smiled at the pale visages of the worldly throng as he passed by.

Among all its bad influences, the black veil had the one desirable effect, of making its wearer a very efficient clergyman. By the aid of his mysterious emblem—for there was no other apparent cause—he became a man of awful power over souls that were in agony for sin. His converts always regarded him with a dread peculiar to themselves, affirming, though but figuratively, that, before he brought them to celestial light, they had been with him behind the black veil. Its gloom, indeed, enabled him to sympathize with all dark affections. Dying sinners cried aloud for Mr. Hooper, and would not yield their breath till he appeared; though ever, as he stooped to whisper consolation, they shuddered at the veiled face so near their own. Such were the terrors of the black veil, even when Death had bared his visage! Strangers came long distances to attend service at his church, with the mere idle purpose of gazing at his figure, because it was forbidden them to behold his face. But many were made to quake ere they departed! Once, during Governor Belcher's administration, Mr. Hooper was appointed to preach the election sermon. Covered with his black veil, he stood before the chief magistrate, the council, and the representatives, and wrought so deep an impression that the legislative measures of that year were characterized by all the gloom and piety of our earliest ancestral sway.

In this manner Mr. Hooper spent a long life, irreproachable in outward act, yet shrouded in dismal suspicions; kind and loving, though unloved, and dimly feared; a man apart from men, shunned in their health and joy, but ever summoned to their aid in mortal anguish. As years wore on, shedding their snows above his sable veil, he acquired a name throughout the New England churches, and they called him Father Hooper. Nearly all his parishioners, who were of mature age when he was settled, had been borne away by many a funeral: he had one congregation in the church, and a more crowded one in the churchyard; and having wrought so late into the evening, and done his work so well, it was now good Father Hooper's turn to rest.

Several persons were visible by the shaded candlelight, in the death chamber of the old clergyman. Natural connections he had none. But there was the decorously grave, though unmoved physician, seeking only to mitigate the last pangs of the patient whom he could not save. There were the deacons, and other eminently pious members of his church. There, also, was the Reverend Mr. Clark, of Westbury, a young and zealous divine, who had ridden in haste to pray by the bedside of the expiring minister. There was the nurse, no hired handmaiden of death, but one whose calm affection had endured thus long in secrecy, in solitude, amid the chill of age, and would not perish, even at the dying hour. Who, but Elizabeth! And there lay the hoary head of good Father Hooper upon the death pillow, with the black veil still swathed about his brow, and reaching down over his face, so that each more difficult gasp of his faint breath caused it to stir. All through life that piece of crape had hung between him and the world: it had separated him from cheerful brotherhood and woman's love, and kept him in that saddest of all prisons, his own heart; and still it lay upon his face, as if to deepen the gloom of his darksome chamber, and shade him from the sunshine of eternity.

For some time previous, his mind had been confused, wavering doubtfully between the past and the present, and hovering forward, as it were, at intervals, into the indistinctness of the world to come. There had been feverish turns, which tossed him from side to side, and wore away what little strength he had. But in his most convulsive struggles, and in the wildest vagaries of his intellect, when no other thought retained its sober influence, he still showed an awful solicitude lest the black veil should slip aside. Even if his bewildered soul could have forgotten, there was a faithful woman at his pillow, who, with averted eyes, would have covered that aged face, which she had last beheld in the comeliness of manhood. At length the death-stricken old man lay quietly in the torpor of mental and bodily exhaustion, with an imperceptible pulse, and breath that grew fainter and fainter, except when a long, deep, and irregular inspiration seemed to prelude the flight of his spirit.

The minister of Westbury approached the bedside.

"Venerable Father Hooper," said he, "the moment of your release is at hand. Are you ready for the lifting of the veil that shuts in time from eternity?"

Father Hooper at first replied merely by a feeble motion of his head; then, apprehensive, perhaps, that his meaning might be doubtful, he exerted himself to speak.

"Yea," said he, in faint accents, "my soul hath a patient weariness until that veil be lifted."

"And is it fitting," resumed the Reverend Mr. Clark, "that a man so given to prayer, of such a blameless example, holy in deed and thought, so far as mortal judgment may pronounce; is it fitting that a father in the church should leave a shadow on his memory, that may seem to blacken a life so pure? I pray you, my venerable brother, let not this thing be! Suffer us to be gladdened by your triumphant aspect as you go to your reward. Before the veil of eternity be lifted, let me cast aside this black veil from your face!"

And thus speaking, the Reverend Mr. Clark bent forward to reveal the mystery of so many years. But, exerting a sudden energy, that made all the beholders stand aghast, Father Hooper snatched both his hands from beneath the bedclothes, and pressed them strongly on the black veil, resolute to struggle, if the minister of Westbury would contend with a dying man.

"Never!" cried the veiled clergyman. "On earth, never!"

"Dark old man!" exclaimed the affrighted minister, "with what horrible crime upon your soul are you now passing to the judgment?"

Father Hooper's breath heaved; it rattled in his throat; but, with a mighty effort, grasping forward with his hands, he caught hold of life, and held it back till he should speak. He even raised himself in bed; and there he sat, shivering with the arms of death around him, while the black veil hung down, awful at that last moment, in the gathered terrors of a lifetime. And yet the faint, sad smile, so often there, now seemed to glimmer from its obscurity, and linger on Father Hooper's lips.

"Why do you tremble at me alone?" cried he, turning his veiled face round the circle of pale spectators. "Tremble also at each other! Have men avoided me, and women shown no pity, and children screamed and fled, only for my black veil? What, but the mystery which it obscurely typifies, has made this piece of crape so awful? When the friend shows his inmost heart to his friend; the lover to his best beloved; when man does not vainly shrink from the eye of his Creator, loathsomely treasuring up the secret of his sin; then deem me a monster, for the symbol beneath which I have lived, and die! I look around me, and, lo! on every visage a Black Veil!"

While his auditors shrank from one another, in mutual affright, Father Hooper fell back upon his pillow, a veiled corpse, with a faint smile lingering on the lips. Still veiled, they laid him in his coffin, and a veiled corpse they bore him to the grave. The grass of many years has sprung up and withered on that grave, the burial stone is moss-grown, and good Mr. Hooper's face is dust; but awful is still the thought that it mouldered beneath the Black Veil!

DISCUSSION QUESTIONS

1. Parson Hooper claims his veil is a symbol, but Poe and others have argued that the moral meaning of the tale is too "obvious," and that a less obvious truth is "insinuated" in relation to the young woman over whose funeral Hooper presides. What evidence in the story supports Poe's claim? Does any evidence in the story support one scholar's argument (Ostrowski 1998) that Hooper's veil might hide the physical signs of syphilis?

2. In early-nineteenth-century America, many people still clung to the puritanical belief that physical illness was a sign of moral failing. Has this attitude toward illness completely vanished, or do we see remnants of it even today? In particular, consider the experiences of patients who may feel stigmatized or marginalized.

3. In the nineteenth century many mental illnesses were lumped together under catchall diagnoses like "monomania" or "moral insanity." Is Hooper a monomaniac? Why or why not?

4. In what ways does the minister maintain control of his own health narrative by wearing the veil and refusing to explain? Compare his attempts to control narrative to Egæus's attempts to control the narrative in Poe's "Berenice." What power lies in one's ability to control his or her own health narrative—or to contribute to the narrative recorded in his or her health care record?

5. At Hooper's deathbed, the doctor, who cannot save Hooper, attempts to mitigate the dying man's suffering. Elizabeth, who acts as his nurse, also provides comfort as Hooper dies. What cultural values are implicit in their desire to ease the patient's suffering at the end of life? What ethical debates surround palliative care in the twenty-first century? How do the concepts of patient autonomy, spirituality, and family dynamics play into those debates?

6. Hawthorne is famous for ambiguity—he leaves meanings and symbols open to multiple interpretations. Do the practices of health care or biomedical science require interpretation and intuition, or do they rely primarily on empirical data? Is there any proper role for imagination in medicine or science?

Dr. Heidegger's Experiment

January 1837, *Knickerbocker,* as "The Fountain of Youth"

In his May 1842 review of Hawthorne's Twice-Told Tales *(Poe 1984, 574), Poe praised Hawthorne as a writer of "true genius" (577). "As Americans, we feel proud of the book," Poe wrote (574). One of the tales Poe expressed an appreciation for was "Dr. Heidegger's Experiment," which he found "exceedingly well imagined and executed with surpassing ability" (575). One can imagine the appeal the story had for Poe. Originally published as "The Fountain of Youth" and collected in* Twice-Told Tales *with the new title that same year, the story focuses on Dr. Heidegger's seemingly innocent experiment to restore youth to a group of four older adults. Poe would have delighted in analyzing the doctor's motives for conducting the experiment and in watching the mystery of the experiment unfold. Clearly, Hawthorne had an enduring interest in the elixir of youth, as he touched upon the idea in the early sketch "Dr. Bullivant" and was exploring the theme even near the end of his life in "Septimius Felton" and "The Dolliver Romance."*

The themes of aging and eternal youth, of course, are highly relevant today as plastic surgery, various cosmetics, and other anti-aging procedures promise their own brand of revitalization and recreation. Our obsession with youthfulness and appearance seems only to have increased throughout the twentieth century and into the twenty-first, even as members of the baby boomer generation enter older adulthood in record numbers. Are Americans willing to undergo experimental procedures for the chance to look and feel younger again? What can we learn about ourselves by studying a group of characters who were willing, almost two centuries ago, to take that chance? How might we apply what we learn to provide a high quality of life and optimal health care to our own aging population?

•

THAT very singular man, old Dr. Heidegger, once invited four venerable friends to meet him in his study. There were three white-bearded gentlemen, Mr. Medbourne, Colonel Killigrew, and Mr. Gascoigne, and a withered gentlewoman, whose name was the Widow Wycherly. They were all melancholy old creatures, who had been unfortunate in life, and whose greatest misfortune it was that they were not long ago in their graves. Mr. Medbourne, in the vigor of his age, had been a prosperous merchant, but had lost his all by a frantic speculation, and was now little better than a mendicant. Colonel Killigrew had wasted his best years, and his health and substance, in the pursuit of sinful pleasures, which had given birth to a brood of pains, such as the gout, and divers other torments of soul and body. Mr. Gascoigne was a ruined politician, a man of evil fame, or at least had been so till time had buried him from the knowledge of the present generation, and made him obscure instead of infamous. As for the Widow Wycherly, tradition tells us that she was a great beauty in her day; but, for a long while past, she had lived in deep seclusion, on account of certain scandalous stories which had prejudiced the gentry of the town against her. It is a circumstance worth mentioning that each of these three old gentlemen, Mr. Medbourne, Colonel Killigrew, and Mr. Gascoigne, were early lovers of the Widow Wycherly, and had once been on the point of cutting each other's throats for her sake. And, before proceeding further, I will merely hint that Dr. Heidegger and all his four guests were sometimes thought to be a little beside themselves—as is not unfrequently the case with old people, when worried either by present troubles or woful recollections.

"My dear old friends," said Dr. Heidegger, motioning them to be seated, "I am desirous of your assistance in one of those little experiments with which I amuse myself here in my study."

If all stories were true, Dr. Heidegger's study must have been a very curious place. It was a dim, old-fashioned chamber, festooned with cobwebs, and be-sprinkled with antique dust. Around the walls stood several oaken bookcases, the lower shelves of which were filled with rows of gigantic folios and black-letter quartos, and the upper with little parchment-covered duodecimos. Over the central bookcase was a bronze bust of Hippocrates, with which, according to some authorities, Dr. Heidegger was accustomed to hold consultations in all difficult cases of his practice. In the obscurest corner of the room stood a tall and narrow oaken closet, with its door ajar, within which doubtfully appeared a skeleton. Between two of the bookcases hung a looking-glass, presenting its high and dusty plate within a tarnished gilt frame. Among many wonderful stories related of this mirror, it was fabled that the spirits of all the doctor's deceased patients dwelt within its verge, and would stare him in the face whenever he looked thitherward. The opposite side of the chamber was ornamented with the full-length portrait of a young lady, arrayed in the faded magnificence of silk, satin, and brocade, and with a visage as faded as her dress. Above half a century ago, Dr. Heidegger had been on the point of marriage with this young lady; but, being affected with some slight disorder, she had swallowed one of her lover's prescriptions, and died on the bridal evening. The greatest curiosity of the study remains to be mentioned; it was a ponderous folio volume, bound in black leather, with massive silver clasps. There were no letters on the back, and nobody could tell the title of the book. But it was well known to be a book of magic; and once, when a chambermaid had lifted it, merely to brush away the dust, the skeleton had rattled in its closet, the picture of the young lady had stepped one foot upon the floor, and several ghastly faces had peeped forth from the mirror; while the brazen head of Hippocrates frowned, and said,—"Forbear!"

Such was Dr. Heidegger's study. On the summer afternoon of our tale a small round table, as black as ebony, stood in the centre of the room, sustaining a cut-glass vase of beautiful form and elaborate workmanship. The sunshine came through the window, between the heavy festoons of two faded damask curtains, and fell directly across this vase; so that a mild splendor was reflected from it on the ashen visages of the five old people who sat around. Four champagne glasses were also on the table.

"My dear old friends," repeated Dr. Heidegger, "may I reckon on your aid in performing an exceedingly curious experiment?"

Now Dr. Heidegger was a very strange old gentleman, whose eccentricity had become the nucleus for a thousand fantastic stories. Some of these fables, to my shame be it spoken, might possibly be traced back to my own veracious self; and if any passages of the present tale should startle the reader's faith, I must be content to bear the stigma of a fiction monger.

When the doctor's four guests heard him talk of his proposed experiment, they anticipated nothing more wonderful than the murder of a mouse in an air pump, or the examination of a cobweb by the microscope, or some similar nonsense, with which he was constantly in the habit of pestering his intimates. But without waiting for a reply, Dr. Heidegger hobbled across the chamber, and returned with the same ponderous folio, bound in black leather, which common report affirmed to be a book of magic. Undoing the silver clasps, he opened the volume, and took from among its black-letter pages a rose, or what was once a rose, though now the green leaves and crimson petals had assumed one brownish hue, and the ancient flower seemed ready to crumble to dust in the doctor's hands.

"This rose," said Dr. Heidegger, with a sigh, "this same withered and crumbling flower, blossomed five and fifty years ago. It was given me by Sylvia Ward, whose portrait hangs yonder; and I meant to wear it in my bosom at our wedding. Five and fifty years it has been treasured between the leaves of this old volume. Now, would you deem it possible that this rose of half a century could ever bloom again?"

"Nonsense!" said the Widow Wycherly, with a peevish toss of her head. "You might as well ask whether an old woman's wrinkled face could ever bloom again."

"See!" answered Dr. Heidegger.

He uncovered the vase, and threw the faded rose into the water which it contained. At first, it lay lightly on the surface of the fluid, appearing to imbibe none of its moisture. Soon, however, a singular change began to be visible. The crushed and dried petals stirred, and assumed a deepening tinge of crimson, as if the flower were reviving from a deathlike slumber; the slender stalk and twigs of foliage became green; and there was the rose of half a century, looking as fresh as when Sylvia Ward had first given it to her lover. It was scarcely full blown; for some of its delicate red leaves curled modestly around its moist bosom, within which two or three dewdrops were sparkling.

"That is certainly a very pretty deception," said the doctor's friends; carelessly, however, for they had witnessed greater miracles at a conjurer's show; "pray how was it effected?"

"Did you never hear of the 'Fountain of Youth?'" asked Dr. Heidegger, "which Ponce de Leon, the Spanish adventurer, went in search of two or three centuries ago?"

"But did Ponce de Leon ever find it?" said the Widow Wycherly.

"No," answered Dr. Heidegger, "for he never sought it in the right place. The famous Fountain of Youth, if I am rightly informed, is situated in the southern part of the Floridian peninsula, not far from Lake Macaco. Its source is overshadowed by several gigantic magnolias, which, though numberless centuries old, have been kept as fresh as violets by the virtues of this wonderful water. An acquaintance of mine, knowing my curiosity in such matters, has sent me what you see in the vase."

"Ahem!" said Colonel Killigrew, who believed not a word of the doctor's story; "and what may be the effect of this fluid on the human frame?"

"You shall judge for yourself, my dear colonel," replied Dr. Heidegger; "and all of you, my respected friends, are welcome to so much of this admirable fluid as may restore to you the bloom of youth. For my own part, having had much trouble in growing old, I am in no hurry to grow young again. With your permission, therefore, I will merely watch the progress of the experiment."

While he spoke, Dr. Heidegger had been filling the four champagne glasses with the water of the Fountain of Youth. It was apparently impregnated with an effervescent gas, for little bubbles were continually ascending from the depths of the glasses, and bursting in silvery spray at the surface. As the liquor diffused a pleasant perfume, the old people doubted not that it possessed cordial and comfortable properties; and though utter sceptics as to its rejuvenescent power, they were inclined to swallow it at once. But Dr. Heidegger besought them to stay a moment.

"Before you drink, my respectable old friends," said he, "it would be well that, with the experience of a lifetime to direct you, you should draw up a few general rules for your guidance, in passing a second time through the perils of youth. Think what a sin and shame it would be, if, with your peculiar advantages, you should not become patterns of virtue and wisdom to all the young people of the age!"

The doctor's four venerable friends made him no answer, except by a feeble and tremulous laugh; so very ridiculous was the idea that, knowing how closely repentance treads behind the steps of error, they should ever go astray again.

"Drink, then," said the doctor, bowing: "I rejoice that I have so well selected the subjects of my experiment."

With palsied hands, they raised the glasses to their lips. The liquor, if it really possessed such virtues as Dr. Heidegger imputed to it, could not have been bestowed on four human beings who needed it more wofully. They looked as if they had never known what youth or pleasure was, but had been the offspring of Nature's dotage, and always the gray, decrepit, sapless, miserable creatures, who now sat stooping round the doctor's table, without life enough in their souls or bodies to be animated even by the prospect of growing young again. They drank off the water, and replaced their glasses on the table.

Assuredly there was an almost immediate improvement in the aspect of the party, not unlike what might have been produced by a glass of generous wine, together with a sudden glow of cheerful sunshine brightening over all their visages at once. There was a healthful suffusion on their cheeks, instead of the ashen hue that had made them look so corpse-like. They gazed at one another, and fancied that some magic power had really begun to smooth away the deep and sad inscriptions which Father Time had been so long engraving on their brows. The Widow Wycherly adjusted her cap, for she felt almost like a woman again.

"Give us more of this wondrous water!" cried they, eagerly. "We are younger—but we are still too old! Quick—give us more!"

"Patience, patience!" quoth Dr. Heidegger, who sat watching the experiment with philosophic coolness. "You have been a long time growing old. Surely, you might be content to grow young in half an hour! But the water is at your service."

Again he filled their glasses with the liquor of youth, enough of which still remained in the vase to turn half the old people in the city to the age of their own grandchildren. While the bubbles were yet sparkling on the brim, the doctor's four guests snatched their glasses from the table, and swallowed the contents at a single gulp. Was it delusion? even while the draught was passing down their throats, it seemed to have wrought a change on their whole systems. Their eyes grew clear and bright; a dark shade deepened among their silvery locks, they sat around the table, three gentlemen of middle age, and a woman, hardly beyond her buxom prime.

"My dear widow, you are charming!" cried Colonel Killigrew, whose eyes had been fixed upon her face, while the shadows of age were flitting from it like darkness from the crimson daybreak.

The fair widow knew, of old, that Colonel Killigrew's compliments were not always measured by sober truth; so she started up and ran to the mirror, still dreading that the ugly visage of an old woman would meet her gaze. Meanwhile, the three gentlemen behaved in such a manner as proved that the water of the Fountain of Youth possessed some intoxicating qualities; unless, indeed, their exhilaration of spirits were merely a lightsome dizziness caused by the sudden removal of the weight of years. Mr. Gascoigne's mind seemed to run on political topics, but whether relating to the past, present, or future could not easily be determined, since the same ideas and phrases have been in vogue these fifty years. Now he rattled forth full-throated sentences about patriotism, national glory, and the people's right; now he muttered some perilous stuff or other, in a sly and doubtful whisper, so cautiously that even his own conscience could scarcely catch the secret; and now, again, he spoke in measured accents, and a deeply deferential tone, as if a royal ear were listening to his well-turned periods. Colonel Killigrew all this time had been trolling forth a jolly bottle song, and ringing his glass in symphony with the chorus, while his eyes wandered toward the buxom figure of the Widow Wycherly. On the other side of the table, Mr. Medbourne was involved in a calculation of dollars and cents, with which was strangely intermingled a project for supplying the East Indies with ice, by harnessing a team of whales to the polar icebergs.

As for the Widow Wycherly, she stood before the mirror courtesying and simpering to her own image, and greeting it as the friend whom she loved better than all the world beside. She thrust her face close to the glass, to see whether some long-remembered wrinkle or crow's foot had indeed vanished. She examined whether the snow had so entirely melted from her hair that the venerable cap could be safely thrown aside. At last, turning briskly away, she came with a sort of dancing step to the table.

"My dear old doctor," cried she, "pray favor me with another glass!"

"Certainly, my dear madam, certainly!" replied the complaisant doctor; "see! I have already filled the glasses."

There, in fact, stood the four glasses, brimful of this wonderful water, the delicate spray of which, as it effervesced from the surface, resembled the tremulous glitter of diamonds. It was now so nearly sunset that the chamber had grown duskier than ever; but a mild and moonlike splendor gleamed from within the vase, and rested alike on the four guests and on the doctor's venerable figure. He sat in a high-backed, elaborately-carved, oaken arm-chair, with a gray dignity of aspect that might have well befitted that very Father Time, whose power had never been disputed, save by this fortunate company. Even while quaffing the third draught of the Fountain of Youth, they were almost awed by the expression of his mysterious visage.

But, the next moment, the exhilarating gush of young life shot through their veins. They were now in the happy prime of youth. Age, with its miserable train of cares and sorrows and diseases, was remembered only as the trouble of a dream, from which they had joyously awoke. The fresh gloss of the soul, so early lost, and without which the world's successive scenes had been but a gallery of faded pictures, again threw its enchantment over all their prospects. They felt like new-created beings in a new-created universe.

"We are young! We are young!" they cried exultingly.

Youth, like the extremity of age, had effaced the strongly-marked characteristics of middle life, and mutually assimilated them all. They were a group of merry youngsters, almost maddened with the exuberant frolicsomeness of their years. The most singular effect of their gayety was an impulse to mock the infirmity and decrepitude of which they had so lately been the victims. They laughed loudly at their old-fashioned attire, the wide-skirted coats and flapped waist-coats of the young men, and the ancient cap and gown of the blooming girl. One limped across the floor like a gouty grandfather; one set a pair of spectacles astride of his nose, and pretended to pore over the black-letter pages of the book of magic; a third seated himself in an arm-chair, and strove to imitate the venerable dignity of Dr. Heidegger. Then all shouted mirthfully, and leaped about the room. The Widow Wycherly—if so fresh a damsel could be called a widow—tripped up to the doctor's chair, with a mischievous merriment in her rosy face.

"Doctor, you dear old soul," cried she, "get up and dance with me!" And then the four young people laughed louder than ever, to think what a queer figure the poor old doctor would cut.

"Pray excuse me," answered the doctor quietly. "I am old and rheumatic, and my dancing days were over long ago. But either of these gay young gentlemen will be glad of so pretty a partner."

"Dance with me, Clara!" cried Colonel Killigrew.

"No, no, I will be her partner!" shouted Mr. Gascoigne.

"She promised me her hand, fifty years ago!" exclaimed Mr. Medbourne.

They all gathered round her. One caught both her hands in his passionate grasp—another threw his arm about her waist—the third buried his hand among the glossy curls that clustered beneath the widow's cap. Blushing, panting, struggling, chiding, laughing, her warm breath fanning each of their faces by turns, she strove to disengage herself, yet still remained in their triple embrace. Never was there a livelier picture of youthful rivalship, with bewitching beauty for the prize. Yet, by a strange deception, owing to the duskiness of the chamber, and the antique dresses which they still wore, the tall mirror is said to have reflected the figures of the three old, gray, withered grandsires, ridiculously contending for the skinny ugliness of a shrivelled grandam.

But they were young: their burning passions proved them so. Inflamed to madness by the coquetry of the girl-widow, who neither granted nor quite withheld her favors, the three rivals began to interchange threatening glances. Still keeping hold of the fair prize, they grappled fiercely at one another's throats. As they struggled to and fro, the table was overturned, and the vase dashed into a thousand fragments. The precious Water of Youth flowed in a bright stream across the floor, moistening the wings of a butterfly, which, grown old in the decline of summer, had alighted there to die. The insect fluttered lightly through the chamber, and settled on the snowy head of Dr. Heidegger.

"Come, come, gentlemen!—come, Madam Wycherly," exclaimed the doctor, "I really must protest against this riot."

They stood still and shivered; for it seemed as if gray Time were calling them back from their sunny youth, far down into the chill and darksome vale of years. They looked at old Dr. Heidegger, who sat in his carved arm-chair, holding the rose of half a century, which he had rescued from among the fragments of the shattered vase. At the motion of his hand, the four rioters resumed their seats; the more readily, because their violent exertions had wearied them, youthful though they were.

"My poor Sylvia's rose!" ejaculated Dr. Heidegger, holding it in the light of the sunset clouds; "it appears to be fading again."

And so it was. Even while the party were looking at it, the flower continued to shrivel up, till it became as dry and fragile as when the doctor had first thrown it into the vase. He shook off the few drops of moisture which clung to its petals.

"I love it as well thus as in its dewy freshness," observed he, pressing the withered rose to his withered lips. While he spoke, the butterfly fluttered down from the doctor's snowy head, and fell upon the floor.

His guests shivered again. A strange chillness, whether of the body or spirit they could not tell, was creeping gradually over them all. They gazed at one another, and fancied that each fleeting moment snatched away a charm, and left a deepening furrow where none had been before. Was it an illusion? Had the changes of a lifetime been crowded into so brief a space, and were they now four aged people, sitting with their old friend, Dr. Heidegger?

"Are we grown old again, so soon?" cried they, dolefully.

In truth they had. The Water of Youth possessed merely a virtue more transient than that of wine. The delirium which it created had effervesced away. Yes! they were old again. With a shuddering impulse, that showed her a woman still, the widow clasped her skinny hands before her face, and wished that the coffin lid were over it, since it could be no longer beautiful.

"Yes, friends, ye are old again," said Dr. Heidegger, "and lo! the Water of Youth is all lavished on the ground. Well—I bemoan it not; for if the fountain gushed at my very doorstep, I would not stoop to bathe my lips in it—no, though its delirium were for years instead of moments. Such is the lesson ye have taught me!"

But the doctor's four friends had taught no such lesson to themselves. They resolved forthwith to make a pilgrimage to Florida, and quaff at morning, noon, and night, from the Fountain of Youth.

NOTE.—In an English review, not long since, I have been accused of plagiarizing the idea of this story from a chapter in one of the novels of Alexandre Dumas. There has undoubtedly been a plagiarism on one side or the other; but as my story was written a good deal more than twenty years ago, and as the novel is of considerably more recent date, I take pleasure in thinking that M. Dumas has done me the honor to appropriate one of the fanciful conceptions of my earlier days. He is heartily welcome to it; nor is it the only instance, by many, in which the French romancer has exercised the privilege of commanding genius by confiscating the intellectual property of less famous people to his own use and behoof.

September, 1860.

DISCUSSION QUESTIONS

1. Compare the setting of Dr. Heidegger's lab to Aylmer's lab in "The Birthmark." How are they alike or different? What does each setting reflect about the practitioner and his work? Then consider what role setting plays in health care today. What impact might the hospital setting have on patients and family members' experiences?

2. How does the story of Sylvia function in this tale? Why does Heidegger share her story with his soon-to-be human subjects, and what information does he withhold? When is it appropriate, if ever, for a biomedical researcher or health care practitioner to withhold information from subjects or patients? How does this tale speak to contemporary issues surrounding full disclosure and informed consent?

3. Consider Dr. Heidegger's level of objectivity, comparing it to that—or lack thereof—exhibited by Aylmer in "The Birthmark" or Dr. Rappaccini in "Rappaccini's Daughter." Nineteenth-century physicians trained in the new clinical medicine believed they should—and could—apply a cold, unbiased eye—what Michel Foucault (1994) called the "clinical gaze"—in research and practice. Is it possible for a researcher or practitioner to be completely objective?

4. Once the Widow Wycherly realizes she cannot be beautiful again permanently, she wishes she were dead. Can you empathize with her response? Why or why not? Do contemporary perceptions of youth and beauty influence twenty-first-century biomedical research and health care practice directly or indirectly? Explain your answers.

5. Write an analysis of a television, magazine, or online commercial for a product advertised as a revolutionary beauty, age, or weight-loss product. Pay particular attention to any pseudoscientific language used to convince the reader on an intellectual level that the product is valid. On what other levels—emotional, et cetera—does the commercial appeal to its audience?

6. Speculate on how the aging baby boomer population will change the delivery of health care in the near future. Do you predict health care providers will feel increasing pressure to offer products and services related to youth and beauty? What ethical dilemmas might arise from those pressures—at both the systemic level and at the provider-patient level?

7. If scientists today claimed to discover the elixir of youth, would you take it? Would you prescribe it to patients? If not, why not? If so, on what terms would you take or prescribe it?

Legends of the Province House
III.
Lady Eleanore's Mantle

December 1838, *United States Magazine and Democratic Review*

Nathaniel Hawthorne and celebrated American poet Henry Wadsworth Long-fellow met as students at Bowdoin College and enjoyed an intermittent friend-ship for the rest of their lives (Hawthorne 1939). Neither man ever met Edgar Allan Poe in person, but Poe accused both of plagiarizing his work. In the case of Longfellow, Poe's charges developed into the now famous "Longfellow War," but Poe's claims against Hawthorne never exploded into such controversy. Poe's assertion that Hawthorne stole lines from his story "William Wilson"—stated in his otherwise positive 1842 review of Twice-Told Tales—*was patently untrue.[1]*

In fact, Hawthorne could have turned the tables on Poe, arguing that he pla-giarized ideas from "Howe's Masquerade" and "Lady Eleanore's Mantle" in "The Masque of the Red Death" (Herndon 2006). Hawthorne made no such accusations, but scholars have noted that Poe read both tales in April 1842 when preparing his review of Twice-Told Tales *and writing "The Masque of the Red Death," which includes imagery similar to Hawthorne's (Herndon 2006; Regan 1970).*

A significant difference between "Lady Eleanore's Mantle" and "The Masque of the Red Death," however, is that the former features a physician character while the latter does not. This highlights a broader difference in their works: Hawthorne, in his illness-themed tales, often features doctors as main characters, while Poe's doctors often remain in the background—absent or fleeting characters. In short, Hawthorne emphasizes the effect of the doctor's presence and his actions, while Poe, with notable exceptions, like the mesmerist tales, represents the doctor as inactive, ineffectual, or outright absent.

In "Lady Eleanore's Mantle," Hawthorne includes a physician character who differs significantly from his other doctor-scientists. Doctor Clarke is still very much present and active here, but he harbors no apparent desire to challenge Mother Nature or experiment with human lives. Although thinly developed, he lacks the malevolence, hubris, and monomania that characterize many of Hawthorne's other doctor-scientists, indicating that Hawthorne had a complex view of doctors and their various relationships with patients and communities.

Readers will enjoy analyzing the characteristics unique to Doctor Clarke and pondering how his role in society differs from that of Hawthorne's deranged monomaniacs. The analysis of his character—as well as of other characters in the tale—will touch upon issues related to empathy, patient-provider relationships, perceptions of mental illness, contagion, and public health. Because "Lady Eleanore's Mantle" is set in pre-revolutionary America, it also gives readers an opportunity to discuss the true history of the smallpox epidemic that struck Boston in the early 1700s and to consider how Hawthorne conceived of certain behaviors—both personal and political—as metaphorically diseased.[2]

•

NOT long after Colonel Shute had assumed the government of Massachusetts Bay, now nearly a hundred and twenty years ago, a young lady of rank and fortune arrived from England, to claim his protection as her guardian. He was her distant relative, but the nearest who had survived the gradual extinction of her family; so that no more eligible shelter could be found for the rich and high-born Lady Eleanore Rochcliffe than within the Province House of a transatlantic colony. The consort of Governor Shute, moreover, had been as a mother to her childhood, and was now anxious to receive her, in the hope that a beautiful young woman would be exposed to infinitely less peril from the primitive society of New England than amid the artifices and corruptions of a court. If either the Governor or his lady had especially consulted their own comfort, they would probably have sought to devolve the responsibility on other hands; since, with some noble and splendid traits of character, Lady Eleanore was remarkable for a harsh, unyielding pride, a haughty consciousness of her hereditary and personal advantages, which made her almost incapable of control. Judging from many traditionary anecdotes, this peculiar temper was hardly less than a monomania; or, if the acts which it inspired were those of a sane person, it seemed due from Providence that pride so sinful should be followed by as severe a retribution. That tinge of the marvellous, which is thrown over so many of these half-forgotten legends, has probably imparted an additional wildness to the strange story of Lady Eleanore Rochcliffe.

The ship in which she came passenger had arrived at Newport, whence Lady Eleanore was conveyed to Boston in the Governor's coach, attended by a small escort of gentlemen on horseback. The ponderous equipage, with its four black horses, attracted much notice as it rumbled through Cornhill, surrounded by the prancing steeds of half a dozen cavaliers, with swords dangling to their stirrups and pistols at their holsters. Through the large glass windows of the coach, as it rolled along, the people could discern the figure of Lady Eleanore, strangely combining an almost queenly stateliness with the grace and beauty of a maiden in her teens. A singular tale had gone abroad among the ladies of the province, that their fair rival was indebted for much of the irresistible charm of her appearance to a certain article of dress—an embroidered mantle—which had been wrought by the most skilful artist in London, and possessed even magical properties of adornment. On the present occasion, however, she owed nothing to the witchery of dress, being clad in a riding habit of velvet, which would have appeared stiff and ungraceful on any other form.

The coachman reined in his four black steeds, and the whole cavalcade came to a pause in front of the contorted iron balustrade that fenced the Province House from the public street. It was an awkward coincidence that the bell of the Old South was just then tolling for a funeral; so that, instead of a gladsome peal with which it was customary to announce the arrival of distinguished strangers, Lady Eleanore Rochcliffe was ushered by a doleful clang, as if calamity had come embodied in her beautiful person.

"A very great disrespect!" exclaimed Captain Langford, an English officer, who had recently brought dispatches to Governor Shute. "The funeral should have been deferred, lest Lady Eleanore's spirits be affected by such a dismal welcome."

"With your pardon, sir," replied Doctor Clarke, a physician, and a famous champion of the popular party, "whatever the heralds may pretend, a dead beggar must have precedence of a living queen. King Death confers high privileges."

These remarks were interchanged while the speakers waited a passage through the crowd, which had gathered on each side of the gateway, leaving an open avenue to the portal of the Province House. A black slave in livery now leaped from behind the coach, and threw open the door; while at the same moment Governor Shute descended the flight of steps from his mansion, to assist Lady Eleanore in alighting. But the Governor's stately approach was anticipated in a manner that excited general astonishment. A pale young man, with his black hair all in disorder, rushed from the throng, and prostrated himself beside the coach, thus offering his person as a footstool for Lady Eleanore Rochcliffe to tread upon. She held back an instant, yet with an expression as if doubting whether the young man were worthy to bear the weight of her footstep, rather than dissatisfied to receive such awful reverence from a fellow-mortal.

"Up, sir," said the Governor, sternly, at the same time lifting his cane over the intruder. "What means the Bedlamite by this freak?"

"Nay," answered Lady Eleanore playfully, but with more scorn than pity in her tone, "your Excellency shall not strike him. When men seek only to be trampled upon, it were a pity to deny them a favor so easily granted—and so well deserved!"

Then, though as lightly as a sunbeam on a cloud, she placed her foot upon the cowering form, and extended her hand to meet that of the Governor. There was a brief interval, during which Lady Eleanore retained this attitude; and never, surely, was there an apter emblem of aristocracy and hereditary pride trampling on human sympathies and the kindred of nature, than these two figures presented at that moment. Yet the spectators were so smitten with her beauty, and so essential did pride seem to the existence of such a creature, that they gave a simultaneous acclamation of applause.

"Who is this insolent young fellow?" inquired Captain Langford, who still remained beside Doctor Clarke. "If he be in his senses, his impertinence demands the bastinado. If mad, Lady Eleanore should be secured from further inconvenience, by his confinement."

"His name is Jervase Helwyse," answered the Doctor; "a youth of no birth or fortune, or other advantages, save the mind and soul that nature gave him; and being secretary to our colonial agent in London, it was his misfortune to meet this Lady Eleanore Rochcliffe. He loved her—and her scorn has driven him mad."

"He was mad so to aspire," observed the English officer.

"It may be so," said Doctor Clarke, frowning as he spoke. "But I tell you, sir, I could well-nigh doubt the justice of the Heaven above us if no signal humiliation overtake this lady, who now treads so haughtily into yonder mansion. She seeks to place herself above the sympathies of our common nature, which envelops all human souls. See, if that nature do not assert its claim over her in some mode that shall bring her level with the lowest!"

"Never!" cried Captain Langford indignantly—"neither in life, nor when they lay her with her ancestors."

Not many days afterwards the Governor gave a ball in honor of Lady Eleanore Rochcliffe. The principal gentry of the colony received invitations, which were distributed to their residences, far and near, by messengers on horseback, bearing missives sealed with all the formality of official dispatches. In obedience to the summons, there was a general gathering of rank, wealth, and beauty; and the wide door of the Province House had seldom given admittance to more numerous and honorable guests than on the evening of Lady Eleanore's ball. Without much extravagance of eulogy, the spectacle might even be termed splendid; for, according to the fashion of the times, the ladies shone in rich silks and satins, outspread over wide-projecting hoops; and the gentlemen glittered in gold embroidery, laid unsparingly upon the purple, or scarlet, or sky-blue velvet, which was the material of their coats and waistcoats. The latter article of dress was of great importance, since it enveloped the wearer's body nearly to the knees, and was perhaps bedizened with the amount of his whole year's income, in golden flowers and foliage. The altered taste of the present day—a taste symbolic of a deep change in the whole system of society—would look upon almost any of those gorgeous figures as ridiculous; although that evening the guests sought their reflections in the pierglasses, and rejoiced to catch their own glitter amid the glittering crowd. What a pity that one of the stately mirrors has not preserved a picture of the scene, which, by the very traits that were so transitory, might have taught us much that would be worth knowing and remembering!

Would, at least, that either painter or mirror could convey to us some faint idea of a garment, already noticed in this legend,—the Lady Eleanore's embroidered mantle—which the gossips whispered was invested with magic properties, so as to lend a new and untried grace to her figure each time that she put it on! Idle fancy as it is, this mysterious mantle has thrown an awe around my image of her, partly from its fabled virtues, and partly because it was the handiwork of a dying woman, and, perchance, owed the fantastic grace of its conception to the delirium of approaching death.

After the ceremonial greetings had been paid, Lady Eleanore Rochcliffe stood apart from the mob of guests, insulating herself within a small and distinguished circle, to whom she accorded a more cordial favor than to the general throng. The waxen torches threw their radiance vividly over the scene, bringing out its brilliant points in strong relief; but she gazed carelessly, and with now and then an expression of weariness or scorn, tempered with such feminine grace that her auditors scarcely perceived the moral deformity of which it was the utterance. She beheld the spectacle not with vulgar ridicule, as disdaining to be pleased with the provincial mockery of a court festival, but with the deeper scorn of one whose spirit held itself too high to participate in the enjoyment of other human souls. Whether or no the recollections of those who saw her that evening were influenced by the strange events with which she was subsequently connected, so it was that her figure ever after recurred to them as marked by something wild and unnatural,—although, at the time, the general whisper was of her exceeding beauty, and of the indescribable charm which her mantle threw around her. Some close observers, indeed, detected a feverish flush and alternate paleness of countenance, with a corresponding flow and revulsion of spirits, and once or twice a painful and helpless betrayal of lassitude, as if she were on the point of sinking to the ground. Then, with a nervous shudder, she seemed to arouse her energies and threw some bright and playful yet half-wicked sarcasm into the conversation. There was so strange a characteristic in her manners and sentiments that it astonished every right-minded listener; till looking in her face, a lurking and incomprehensible glance and smile perplexed them with doubts both as to her seriousness and sanity. Gradually, Lady Eleanore Rochcliffe's circle grew smaller, till only four gentlemen remained in it. These were Captain Langford, the English officer before mentioned; a Virginian planter, who had come to Massachusetts on some political errand; a young Episcopal clergyman, the grandson of a British earl; and, lastly, the private secretary of Governor Shute, whose obsequiousness had won a sort of tolerance from Lady Eleanore.

At different periods of the evening the liveried servants of the Province House passed among the guests, bearing huge trays of refreshments and French and Spanish wines. Lady Eleanore Rochcliffe, who refused to wet her beautiful lips even with a bubble of Champagne, had sunk back into a large damask chair, apparently overwearied either with the excitement of the scene or its tedium, and while, for an instant, she was unconscious of voices, laughter, and music, a young man stole forward, and knelt down at her feet. He bore a salver in his hand, on which was a chased silver goblet, filled to the brim with wine, which he offered as reverentially as to a crowned queen, or rather with the awful devotion of a priest doing sacrifice to his idol. Conscious that someone touched her robe, Lady Eleanore started, and unclosed her eyes upon the pale, wild features and dishevelled hair of Jervase Helwyse.

"Why do you haunt me thus?" said she, in a languid tone, but with a kindlier feeling than she ordinarily permitted herself to express. "They tell me that I have done you harm."

"Heaven knows if that be so," replied the young man solemnly. "But, Lady Eleanore, in requital of that harm, if such there be, and for your own earthly and heavenly welfare, I pray you to take one sip of this holy wine, and then to pass the goblet round among the guests. And this shall be a symbol that you have not sought to withdraw yourself from the chain of human sympathies—which whoso would shake off must keep company with fallen angels."

"Where has this mad fellow stolen that sacramental vessel?" exclaimed the Episcopal clergyman.

This question drew the notice of the guests to the silver cup, which was recognized as appertaining to the communion plate of the Old South Church; and, for aught that could be known, it was brimming over with the consecrated wine.

"Perhaps it is poisoned," half whispered the Governor's secretary.

"Pour it down the villain's throat!" cried the Virginian fiercely.

"Turn him out of the house!" cried Captain Langford, seizing Jervase Helwyse so roughly by the shoulder that the sacramental cup was overturned, and its contents sprinkled upon Lady Eleanore's mantle. "Whether knave, fool, or Bedlamite, it is intolerable that the fellow should go at large."

"Pray, gentlemen, do my poor admirer no harm," said Lady Eleanore, with a faint and weary smile. "Take him out of my sight, if such be your pleasure; for I can find in my heart to do nothing but laugh at him; whereas, in all decency and conscience, it would become me to weep for the mischief I have wrought!"

But while the by-standers were attempting to lead away the unfortunate young man, he broke from them, and with a wild, impassioned earnestness, offered a new and equally strange petition to Lady Eleanore. It was no other than that she should throw off the mantle, which, while he pressed the silver cup of wine upon her, she had drawn more closely around her form, so as almost to shroud herself within it.

"Cast it from you!" exclaimed Jervase Helwyse, clasping his hands in an agony of entreaty. "It may not yet be too late! Give the accursed garment to the flames!"

But Lady Eleanore, with a laugh of scorn, drew the rich folds of the embroidered mantle over her head, in such a fashion as to give a completely new aspect to her beautiful face, which—half hidden, half revealed—seemed to belong to some being of mysterious character and purposes.

"Farewell, Jervase Helwyse!" said she. "Keep my image in your remembrance, as you behold it now."

"Alas, lady!" he replied, in a tone no longer wild, but sad as a funeral bell. "We must meet shortly, when your face may wear another aspect—and that shall be the image that must abide within me."

He made no more resistance to the violent efforts of the gentlemen and servants, who almost dragged him out of the apartment, and dismissed him roughly from the iron gate of the Province House. Captain Langford, who had been very active in this affair, was returning to the presence of Lady Eleanore Rochcliffe, when he encountered the physician, Doctor Clarke, with whom he had held some casual talk on the day of her arrival. The Doctor stood apart, separated from Lady Eleanore by the width of the room, but eying her with such keen sagacity that Captain Langford involuntarily gave him credit for the discovery of some deep secret.

"You appear to be smitten, after all, with the charms of this queenly maiden," said he, hoping thus to draw forth the physician's hidden knowledge.

"God forbid!" answered Doctor Clarke, with a grave smile; "and if you be wise you will put up the same prayer for yourself. Wo to those who shall be smitten by this beautiful Lady Eleanore! But yonder stands the Governor—and I have a word or two for his private ear. Good night!"

He accordingly advanced to Governor Shute, and addressed him in so low a tone that none of the by-standers could catch a word of what he said, although the sudden change of his Excellency's hitherto cheerful visage betokened that the communication could be of no agreeable import. A very few moments afterwards it was announced to the guests that an unforeseen circumstance rendered it necessary to put a premature close to the festival.

The ball at the Province House supplied a topic of conversation for the colonial metropolis for some days after its occurrence, and might still longer have been the general theme, only that a subject of all-engrossing interest thrust it, for a time, from the public recollection. This was the appearance of a dreadful epidemic, which, in that age and long before and afterwards, was wont to slay its hundreds and thousands on both sides of the Atlantic. On the occasion of which we speak, it was distinguished by a peculiar virulence, insomuch that it has left its traces—its pit-marks, to use an appropriate figure—on the history of the country, the affairs of which were thrown into confusion by its ravages. At first, unlike its ordinary course, the disease seemed to confine itself to the higher circles of society, selecting its victims from among the proud, the well-born, and the wealthy, entering unabashed into stately chambers, and lying down with the slumberers in silken beds. Some of the most distinguished guests of the Province House—even those whom the haughty Lady Eleanore Rochcliffe had deemed not unworthy of her favor—were stricken by this fatal scourge. It was noticed, with an ungenerous bitterness of feeling, that the four gentlemen—the Virginian, the British officer, the young clergyman, and the Governor's secretary—who had been her most devoted attendants on the evening of the ball, were the foremost on whom the plague stroke fell. But the disease, pursuing its onward progress, soon ceased to be exclusively a prerogative of aristocracy. Its red brand was no longer conferred like a noble's star, or an order of knighthood. It threaded its way through the narrow and crooked streets, and entered the low, mean, darksome dwellings, and laid its hand of death upon the artisans and laboring classes of the town. It compelled rich and poor to feel themselves brethren then; and stalking to and fro across the Three Hills, with a fierceness which made it almost a new pestilence, there was that mighty conqueror—that scourge and horror of our forefathers—the Small-Pox!

We cannot estimate the affright which this plague inspired of yore, by contemplating it as the fangless monster of the present day. We must remember, rather, with what awe we watched the gigantic footsteps of the Asiatic cholera, striding from shore to shore of the Atlantic, and marching like destiny upon cities far remote which flight had already half depopulated. There is no other fear so horrible and unhumanizing as that which makes man dread to breathe heaven's vital air lest it be poison, or to grasp the hand of a brother or friend lest the gripe of the pestilence should clutch him. Such was the dismay that now followed in the track of the disease, or ran before it throughout the town. Graves were hastily dug, and the pestilential relics as hastily covered, because the dead were enemies of the living, and strove to draw them headlong, as it were, into their own dismal pit. The public councils were suspended, as if mortal wisdom might relinquish its devices, now that an unearthly usurper had found his way into the ruler's mansion. Had an enemy's fleet been hovering on the coast, or his armies trampling on our soil, the people would probably have committed their defence to that same direful conqueror who had wrought their own calamity, and would permit no interference with his sway. This conqueror had a symbol of his triumphs. It was a blood-red flag, that fluttered in the tainted air, over the door of every dwelling into which the Small-Pox had entered.

Such a banner was long since waving over the portal of the Province House; for thence, as was proved by tracking its footsteps back, had all this dreadful mischief issued. It had been traced back to a lady's luxurious chamber—to the proudest of the proud—to her that was so delicate, and hardly owned herself of earthly mould—to the haughty one, who took her stand above human sympathies—to Lady Eleanore! There remained no room for doubt that the contagion had lurked in that gorgeous mantle, which threw so strange a grace around her at the festival. Its fantastic splendor had been conceived in the delirious brain of a woman on her death-bed, and was the last toil of her stiffening fingers, which had interwoven fate and misery with its golden threads. This dark tale, whispered at first, was now bruited far and wide. The people raved against the Lady Eleanore, and cried out that her pride and scorn had evoked a fiend, and that, between them both, this monstrous evil had been born. At times, their rage and despair took the semblance of grinning mirth; and whenever the red flag of the pestilence was hoisted over another and yet another door, they clapped their hands and shouted through the streets, in bitter mockery: "Behold a new triumph for the Lady Eleanore!"

One day, in the midst of these dismal times, a wild figure approached the portal of the Province House, and folding his arms, stood contemplating the scarlet banner which a passing breeze shook fitfully, as if to fling abroad the contagion that it typified. At length, climbing one of the pillars by means of the iron balustrade, he took down the flag and entered the mansion, waving it above his head. At the foot of the staircase he met the Governor, booted and spurred, with his cloak drawn around him, evidently on the point of setting forth upon a journey.

"Wretched lunatic, what do you seek here?" exclaimed Shute, extending his cane to guard himself from contact. "There is nothing here but Death. Back—or you will meet him!"

"Death will not touch me, the banner-bearer of the pestilence!" cried Jervase Helwyse, shaking the red flag aloft. "Death, and the Pestilence, who wears the aspect of the Lady Eleanore, will walk through the streets to-night, and I must march before them with this banner!"

"Why do I waste words on the fellow?" muttered the Governor, drawing his cloak across his mouth. "What matters his miserable life, when none of us are sure of twelve hours' breath? On, fool, to your own destruction!"

He made way for Jervase Helwyse, who immediately ascended the staircase, but, on the first landing-place, was arrested by the firm grasp of a hand upon his shoulder. Looking fiercely up, with a madman's impulse to struggle with and rend asunder his opponent, he found himself powerless beneath a calm, stern eye, which possessed the mysterious property of quelling frenzy at its height. The person whom he had now encountered was the physician, Doctor Clarke, the duties of whose sad profession had led him to the Province House, where he was an infrequent guest in more prosperous times.

"Young man, what is your purpose?" demanded he.

"I seek the Lady Eleanore," answered Jervase Helwyse, submissively.

"All have fled from her," said the physician. "Why do you seek her now? I tell you, youth, her nurse fell death-stricken on the threshold of that fatal chamber. Know ye not, that never came such a curse to our shores as this lovely Lady Eleanore?—that her breath has filled the air with poison?—that she has shaken pestilence and death upon the land, from the folds of her accursed mantle?"

"Let me look upon her!" rejoined the mad youth, more wildly. "Let me behold her, in her awful beauty, clad in the regal garments of the pestilence! She and Death sit on a throne together. Let me kneel down before them!"

"Poor youth!" said Doctor Clarke; and, moved by a deep sense of human weakness, a smile of caustic humor curled his lip even then. "Wilt thou still worship the destroyer and surround her image with fantasies the more magnificent, the more evil she has wrought? Thus man doth ever to his tyrants. Approach, then! Madness, as I have noted, has that good efficacy, that it will guard you from contagion—and perchance its own cure may be found in yonder chamber."

Ascending another flight of stairs, he threw open a door and signed to Jervase Helwyse that he should enter. The poor lunatic, it seems probable, had cherished a delusion that his haughty mistress sat in state, unharmed herself by the pestilential influence, which, as by enchantment, she scattered round about her. He dreamed, no doubt, that her beauty was not dimmed, but brightened into superhuman splendor. With such anticipations, he stole reverentially to the door at which the physician stood, but paused upon the threshold gazing fearfully into the gloom of the darkened chamber.

"Where is the Lady Eleanore?" whispered he.

"Call her," replied the physician.

"Lady Eleanore!—Princess!—Queen of Death!" cried Jervase Helwyse, advancing three steps into the chamber. "She is not here! There, on yonder table, I behold the sparkle of a diamond which once she wore upon her bosom. There"—and he shuddered—"there hangs her mantle, on which a dead woman embroidered a spell of dreadful potency. But where is the Lady Eleanore?"

Something stirred within the silken curtains of a canopied bed; and a low moan was uttered, which, listening intently, Jervase Helwyse began to distinguish as a woman's voice, complaining dolefully of thirst. He fancied, even, that he recognized its tones.

"My throat!—my throat is scorched," murmured the voice. "A drop of water!"

"What thing art thou?" said the brain-stricken youth, drawing near the bed and tearing asunder its curtains. "Whose voice hast thou stolen for thy murmurs and miserable petitions, as if Lady Eleanore could be conscious of mortal infirmity? Fie! Heap of diseased mortality, why lurkest thou in my lady's chamber?"

"O Jervase Helwyse," said the voice—and as it spoke the figure contorted itself, struggling to hide its blasted face—"look not now on the woman you once loved! The curse of Heaven hath stricken me, because I would not call man my brother, nor woman sister. I wrapped myself in PRIDE as in a MANTLE, and scorned the sympathies of nature; and therefore has nature made this wretched body the medium of a dreadful sympathy. You are avenged—they are all avenged—Nature is avenged—for I am Eleanore Rochcliffe!"

The malice of his mental disease, the bitterness lurking at the bottom of his heart, mad as he was, for a blighted and ruined life, and love that had been paid with cruel scorn, awoke within the breast of Jervase Helwyse. He shook his finger at the wretched girl, and the chamber echoed, the curtains of the bed were shaken, with his outburst of insane merriment.

"Another triumph for the Lady Eleanore!" he cried. "All have been her victims! Who so worthy to be the final victim as herself?"

Impelled by some new fantasy of his crazed intellect, he snatched the fatal mantle and rushed from the chamber and the house. That night a procession passed, by torchlight, through the streets, bearing in the midst the figure of a woman, enveloped with a richly embroidered mantle; while in advance stalked Jervase Helwyse, waving the red flag of the pestilence. Arriving opposite the Province House, the mob burned the effigy, and a strong wind came and swept away the ashes. It was said that, from that very hour, the pestilence abated, as if its sway had some mysterious connection, from the first plague stroke to the last, with Lady Eleanore's Mantle. A remarkable uncertainty broods over that unhappy lady's fate. There is a belief, however, that in a certain chamber of this mansion a female form may sometimes be duskily discerned, shrinking into the darkest corner and muffling her face within an embroidered mantle. Supposing the legend true, can this be other than the once proud Lady Eleanore?

DISCUSSION QUESTIONS

1. Write down three adjectives to describe each of the following characters: Lady Eleanore, Jervase Helwyse, and Doctor Clarke. What defining traits does each character have? Does Doctor Clarke's impression of Lady Eleanore affect the care he provides her in the tale? Explain your answer.

2. If Lady Eleanore's illness is a punishment for her crimes against "the sympathies of nature," how might we explain Jervase Helwyse's mental illness? Is his illness a punishment for a "sin" he has committed? How would Doctor Clarke answer this question? As a caregiver, how would you respond to a patient who believed his or her illness was the result of personal "sin" or perhaps "karma"? Consider the role therapeutic communication plays in helping providers respond compassionately and nonjudgmentally to patients with such concerns. What other types of communication play a significant role in the delivery of compassionate care? Consider reading Sayantani DasGupta's (2008) "Narrative Humility" and including the essay in your discussion.

3. Consider the theme of power in the tale. Who has power, and who is powerless? In what ways does the arrival of smallpox shift these power structures? If you are a health care student or practitioner, what power structures do you operate within every day? Choose one and write a one-page reflection on how this power structure either positively or negatively influences patient safety and outcomes. Share highlights from your reflection with the group.

4. Doctor Clarke has a strong voice in the tale that helps shape the reader's opinions about other characters. If you were a patient, how much would it matter to you who got to "tell your tale"? How much of a voice are patients given in the typical health care encounter today? Does the typical illness narrative "belong" to the patient or to the provider or to both? In what ways do the movements in patient-, family-, and relationship-centered care speak to this question?

5. Compare Doctor Clarke with other doctor-scientists in Hawthorne's tales. How, for example, is he like or unlike Doctor Bullivant, another pre-revolutionary doctor? In what ways does each character reflect political and historical realities of his respective time period? In what ways do all health care practitioners reflect the social or political ideologies of their time—and in what ways do those ideologies affect biomedical research and patient care? Provide specific examples.

The Birthmark

March 1843, *Pioneer*

Perhaps it goes without saying that Hawthorne was not, in his private life, a very progressive man when compared to reform-minded contemporaries like Ralph Waldo Emerson and Margaret Fuller. While other romantics championed abolitionist and feminist causes, Hawthorne notably did not. And yet he produced several of the most pro-feminist literary works of nineteenth-century America (Millington 2014), indicting medical and scientific practices that objectified, dehumanized, and ultimately destroyed women.

Was he influenced by personal experiences—having seen firsthand how experimental pseudoscientific practices had negatively affected—and never healed—his mother and wife? Or was he, as literary scholars Nina Baym (1982; 2005), Judith Fetterly (1978), and others have argued, criticizing male obsession and scientific ambition—traits promoted by a culture that offered rigid definitions of masculinity and femininity (Millington 2014)? Did he as a man resent this rigidity because he as a writer was working in a profession that valued contemplation, intuition, and imagination over action and aggression (Elbert 2004a)? Regardless of his motivations, in "The Birthmark" Hawthorne created a powerful tale that allegorically addresses the plight of women living during the era of the Cult of True Womanhood.

The Cult of True Womanhood, which labeled women as biologically inferior to men, warned women not to take on too much stress or overstimulate themselves intellectually, since these activities might cause a mental breakdown or compromise reproductive abilities. When treating women for physical or mental ailments, many of which were lumped under the diagnosis of "hysteria," male doctors "focused on the pelvis" (Wendland 2006, 32). Although the ancient Greek physician and anatomist Galen had long ago noted that the uterus was a stationary organ, early-nineteenth-century physicians continued to believe in the ancient Egyptian and Greek theory of "wandering uterus," which proposed that the uterus would float around the body, causing various illnesses. A nineteenth-century woman who suffered from menstrual cramps or sometimes menopause was offered a surgical option (hysterectomy), with a 40 percent mortality rate, or she could opt to take Lydia Pinkham's vegetable tonic designed to help her avoid such drastic measures (Harvard University Library 2015).

As a result of the widely accepted—albeit misunderstood—connection between the reproductive system and women's health, clitoridectomies were later performed to address masturbation before it could lead to "a series of disasters progressing through insomnia, exhaustion, neurasthenia, epilepsy, moral insanity, insanity, convulsions, melancholia and paralysis, to eventual coma and death" (Studd 2007, 674). Bilateral oophorectomies were indicated for a number of ailments, from nymphomania to depression to epilepsy (32). Newer surgical techniques were also often pioneered on women, particularly the enslaved, so that surgeons could refine their techniques for wider use (Wendland 2006, 34). Thus, women of all classes and stations were vulnerable to medical experimentation.

Pioneering women doctors like Elizabeth Blackwell, who began to practice a decade after the publication of "The Birthmark," viewed these techniques as exploitative. However, female physicians were rare in a medical world where women were "typically the object of scientific knowledge and surgical technique, not its practitioner" (Wendland 2006, 34). In fact, at the time the term "female physician" commonly referred to a woman who performed abortions (Sahli 1982).

While we have certainly come a long way since the Cult of True Womanhood, "The Birthmark" remains relevant to the discussion of women's health—and to broader discussions of bioethics in medicine and research. The tale also challenges us to discuss body image and appearance, to consider what role culture plays in shaping the way we view ourselves and each other—and to what dangerous lengths some individuals go to get one step closer to physical perfection.[1]

•

In the latter part of the last century there lived a man of science, an eminent proficient in every branch of natural philosophy, who not long before our story opens had made experience of a spiritual affinity more attractive than any chemical one. He had left his laboratory to the care of an assistant, cleared his fine countenance from the furnace smoke, washed the stain of acids from his fingers, and persuaded a beautiful woman to become his wife. In those days when the comparatively recent discovery of electricity and other kindred mysteries of Nature seemed to open paths into the region of miracle, it was not unusual for the love of science to rival the love of woman in its depth and absorbing energy. The higher intellect, the imagination, the spirit, and even the heart might all find their congenial aliment in pursuits which, as some of their ardent votaries believed, would ascend from one step of powerful intelligence to another, until the philosopher should lay his hand on the secret of creative force and perhaps make new worlds for himself. We know not whether Aylmer possessed this degree of faith in man's ultimate control over Nature. He had devoted himself, however, too unreservedly to scientific studies ever to be weaned from them by any second passion. His love for his young wife might prove the stronger of the two; but it could only be by intertwining itself with his love of science, and uniting the strength of the latter to its own.

Such a union accordingly took place, and was attended with truly remarkable consequences and a deeply impressive moral. One day, very soon after their marriage, Aylmer sat gazing at his wife with a trouble in his countenance that grew stronger until he spoke.

"Georgiana," said he, "has it never occurred to you that the mark upon your cheek might be removed?"

"No, indeed," said she, smiling; but perceiving the seriousness of his manner, she blushed deeply. "To tell you the truth, it has been so often called a charm that I was simple enough to imagine it might be so."

"Ah, upon another face, perhaps it might," replied her husband; "but never on yours. No, dearest Georgiana, you came so nearly perfect from the hand of Nature that this slightest possible defect, which we hesitate whether to term a defect or a beauty, shocks me, as being the visible mark of earthly imperfection."

"Shocks you, my husband!" cried Georgiana, deeply hurt; at first reddening with momentary anger, but then bursting into tears. "Then why did you take me from my mother's side? You cannot love what shocks you!"

To explain this conversation it must be mentioned that in the centre of Georgiana's left cheek, there was a singular mark, deeply interwoven, as it were, with the texture and substance of her face. In the usual state of her complexion—a healthy though delicate bloom—the mark wore a tint of deeper crimson, which imperfectly defined its shape amid the surrounding rosiness. When she blushed it gradually became more indistinct, and finally vanished amid the triumphant rush of blood that bathed the whole cheek with its brilliant glow. But if any shifting emotion caused her to turn pale there was the mark again, a crimson stain upon the snow, in what Aylmer sometimes deemed an almost fearful distinctness. Its shape bore not a little similarity to the human hand, though of the smallest pigmy size. Georgiana's lovers were wont to say, that some fairy at her birth hour had laid her tiny hand upon the infant's cheek, and left this impress there in token of the magic endowments that were to give her such sway over all hearts. Many a desperate swain would have risked life for the privilege of pressing his lips to the mysterious hand. It must not be concealed, however, that the impression wrought by this fairy sign manual varied exceedingly, according to the difference of temperament in the beholders. Some fastidious persons—but they were exclusively of her own sex—affirmed that the bloody hand, as they chose to call it, quite destroyed the effect of Georgiana's beauty, and rendered her countenance even hideous. But it would be as reasonable to say that one of those small blue stains which sometimes occur in the purest statuary marble would convert the Eve of Powers to a monster. Masculine observers, if the birthmark did not heighten their admiration, contented themselves with wishing it away, that the world might possess one living specimen of ideal loveliness, without the semblance of a flaw. After his marriage,—for he thought little or nothing of the matter before—Aylmer discovered that this was the case with himself.

Had she been less beautiful,—if Envy's self could have found aught else to sneer at,—he might have felt his affection heightened by the prettiness of this mimic hand, now vaguely portrayed, now lost, now stealing forth again and glimmering to and fro with every pulse of emotion that throbbed within her heart; but, seeing her otherwise so perfect, he found this one defect grow more and more intolerable with every moment of their united lives. It was the fatal flaw of humanity which Nature, in one shape or another, stamps ineffaceably on all her productions, either to imply that they are temporary and finite, or that their perfection must be wrought by toil and pain. The crimson hand expressed the ineludible gripe in which mortality clutches the highest and purest of earthly mould, degrading them into kindred with the lowest, and even with the very brutes, like whom their visible frames return to dust. In this manner, selecting it as the symbol of his wife's liability to sin, sorrow, decay, and death, Aylmer's sombre imagination was not long in rendering the birthmark a frightful object, causing him more trouble and horror than ever Georgiana's beauty, whether of soul or sense, had given him delight.

At all the seasons which should have been their happiest, he invariably and without intending it, nay, in spite of a purpose to the contrary, reverted to this one disastrous topic. Trifling as it at first appeared, it so connected itself with innumerable trains of thought and modes of feeling that it became the central point of all. With the morning twilight, Aylmer opened his eyes upon his wife's face and recognized the symbol of imperfection; and when they sat together at the evening hearth his eyes wandered stealthily to her cheek, and beheld, flickering with the blaze of the wood fire, the spectral hand that wrote mortality where he would fain have worshipped. Georgiana soon learned to shudder at his gaze. It needed but a glance with the peculiar expression that his face often wore to change the roses of her cheek into a deathlike paleness, amid which the crimson hand was brought strongly out, like a bass-relief of ruby on the whitest marble.

Late one night when the lights were growing dim, so as hardly to betray the stain on the poor wife's cheek, she herself, for the first time, voluntarily took up the subject.

"Do you remember, my dear Aylmer," said she, with a feeble attempt at a smile, "have you any recollection of a dream last night about this odious hand?"

"None! none whatever!" replied Aylmer, starting; but then he added in a dry, cold tone, affected for the sake of concealing the real depth of his emotion, "I might well dream of it; for before I fell asleep it had taken a pretty firm hold of my fancy."

"And you did dream of it?" continued Georgiana, hastily; for she dreaded lest a gush of tears should interrupt what she had to say. "A terrible dream! I wonder that you can forget it. Is it possible to forget this one expression?—'It is in her heart now; we must have it out!' Reflect, my husband; for by all means I would have you recall that dream."

The mind is in a sad state when Sleep, the all-involving, cannot confine her spectres within the dim region of her sway, but suffers them to break forth, affrighting this actual life with secrets that perchance belong to a deeper one. Aylmer now remembered his dream. He had fancied himself with his servant Aminadab, attempting an operation for the removal of the birthmark; but the deeper went the knife, the deeper sank the hand, until at length its tiny grasp appeared to have caught hold of Georgiana's heart; whence, however, her husband was inexorably resolved to cut or wrench it away.

When the dream had shaped itself perfectly in his memory, Aylmer sat in his wife's presence with a guilty feeling. Truth often finds its way to the mind close muffled in robes of sleep, and then speaks with uncompromising directness of matters in regard to which we practise an unconscious self-deception during our waking moments. Until now he had not been aware of the tyrannizing influence acquired by one idea over his mind, and of the lengths which he might find in his heart to go for the sake of giving himself peace.

"Aylmer," resumed Georgiana, solemnly, "I know not what may be the cost to both of us to rid me of this fatal birthmark. Perhaps its removal may cause cureless deformity; or it may be the stain goes as deep as life itself. Again: do we know that there is a possibility, on any terms, of unclasping the firm gripe of this little hand which was laid upon me before I came into the world?"

"Dearest Georgiana, I have spent much thought upon the subject," hastily interrupted Aylmer. "I am convinced of the perfect practicability of its removal."

"If there be the remotest possibility of it," continued Georgiana, "let the attempt be made at what ever risk. Danger is nothing to me; for life, while this hateful mark makes me the object of your horror and disgust,—life is a burthen which I would fling down with joy. Either remove this dreadful hand, or take my wretched life! You have deep science. All the world bears witness of it. You have achieved great wonders. Cannot you remove this little, little mark, which I cover with the tips of two small fingers! Is this beyond your power, for the sake of your own peace, and to save your poor wife from madness?"

"Noblest, dearest, tenderest wife!" cried Aylmer, rapturously, "doubt not my power. I have already given this matter the deepest thought—thought which might almost have enlightened me to create a being less perfect than yourself. Georgiana, you have led me deeper than ever into the heart of science. I feel myself fully competent to render this dear cheek as faultless as its fellow; and then, most beloved, what will be my triumph when I shall have corrected what Nature left imperfect in her fairest work! Even Pygmalion, when his sculptured woman assumed life, felt not greater ecstasy than mine will be."

"It is resolved, then," said Georgiana, faintly smiling. "And, Aylmer, spare me not, though you should find the birthmark take refuge in my heart at last."

Her husband tenderly kissed her cheek—her right cheek—not that which bore the impress of the crimson hand.

The next day, Aylmer apprised his wife of a plan that he had formed, whereby he might have opportunity for the intense thought and constant watchfulness which the proposed operation would require; while Georgiana, likewise, would enjoy the perfect repose essential to its success. They were to seclude themselves in the extensive apartments occupied by Aylmer as a laboratory, and where, during his toilsome youth, he had made discoveries in the elemental powers of Nature that had roused the admiration of all the learned societies in Europe. Seated calmly in this laboratory, the pale philosopher had investigated the secrets of the highest cloud region, and of the profoundest mines; he had satisfied himself of the causes that kindled and kept alive the fires of the volcano; and had explained the mystery of fountains, and how it is that they gush forth, some so bright and pure, and others with such rich medicinal virtues, from the dark bosom of the earth. Here, too, at an earlier period, he had studied the wonders of the human frame, and attempted to fathom the very process by which Nature assimilates all her precious influences from earth and air, and from the spiritual world, to create and foster man, her masterpiece. The latter pursuit, however, Aylmer had long laid aside, in unwilling recognition of the truth—against which all seekers sooner or later stumble—that our great creative Mother, while she amuses us with apparently working in the broadest sunshine, is yet severely careful to keep her own secrets, and, in spite of her pretended openness, shows us nothing but results. She permits us, indeed, to mar, but seldom to mend, and, like a jealous patentee, on no account to make. Now, however, Aylmer resumed these half-forgotten investigations; not, of course, with such hopes or wishes as first suggested them; but because they involved much physiological truth, and lay in the path of his proposed scheme for the treatment of Georgiana.

As he led her over the threshold of the laboratory, Georgiana was cold and tremulous. Aylmer looked cheerfully into her face, with intent to reassure her, but was so startled with the intense glow of the birthmark upon the whiteness of her cheek that he could not restrain a strong convulsive shudder. His wife fainted.

"Aminadab! Aminadab!" shouted Aylmer, stamping violently on the floor.

Forthwith, there issued from an inner apartment a man of low stature, but bulky frame, with shaggy hair hanging about his visage, which was grimed with the vapors of the furnace. This personage had been Aylmer's underworker during his whole scientific career, and was admirably fitted for that office by his great mechanical readiness, and the skill with which, while incapable of comprehending a single principle, he executed all the practical details of his master's experiments. With his vast strength, his shaggy hair, his smoky aspect, and the indescribable earthiness that incrusted him, he seemed to represent man's physical nature; while Aylmer's slender figure, and pale, intellectual face, were no less apt a type of the spiritual element.

"Throw open the door of the boudoir, Aminadab," said Aylmer, "and burn a pastil."

"Yes, master," answered Aminadab, looking intently at the lifeless form of Georgiana; and then he muttered to himself: "If she were my wife, I'd never part with that birthmark."

When Georgiana recovered consciousness she found herself breathing an atmosphere of penetrating fragrance, the gentle potency of which had recalled her from her deathlike faintness. The scene around her looked like enchantment. Aylmer had converted those smoky, dingy, sombre rooms, where he had spent his brightest years in recondite pursuits, into a series of beautiful apartments not unfit to be the secluded abode of a lovely woman. The walls were hung with gorgeous curtains, which imparted the combination of grandeur and grace that no other species of adornment can achieve; and as they fell from the ceiling to the floor, their rich and ponderous folds, concealing all angles and straight lines, appeared to shut in the scene from infinite space. For aught Georgiana knew, it might be a pavilion among the clouds. And Aylmer, excluding the sunshine, which would have interfered with his chemical processes, had supplied its place with perfumed lamps, emitting flames of various hue, but all uniting in a soft, impurpled radiance. He now knelt by his wife's side, watching her earnestly, but without alarm; for he was confident in his science, and felt that he could draw a magic circle round her within which no evil might intrude.

"Where am I? Ah, I remember," said Georgiana, faintly; and she placed her hand over her cheek, to hide the terrible mark from her husband's eyes.

"Fear not, dearest!" exclaimed he. "Do not shrink from me! Believe me, Georgiana, I even rejoice in this single imperfection, since it will be such a rapture to remove it."

"Oh, spare me!" sadly replied his wife. "Pray do not look at it again. I never can forget that convulsive shudder."

In order to soothe Georgiana, and, as it were, to release her mind from the burden of actual things, Aylmer now put in practice some of the light and playful secrets which science had taught him among its profounder lore. Airy figures, absolutely bodiless ideas, and forms of unsubstantial beauty came and danced before her, imprinting their momentary footsteps on beams of light. Though she had some indistinct idea of the method of these optical phenomena, still the illusion was almost perfect enough to warrant the belief that her husband possessed sway over the spiritual world. Then again, when she felt a wish to look forth from her seclusion, immediately, as if her thoughts were answered, the procession of external existence flitted across a screen. The scenery and the figures of actual life were perfectly represented, but with that bewitching, yet indescribable difference which always makes a picture, an image, or a shadow so much more attractive than the original. When wearied of this, Aylmer bade her cast her eyes upon a vessel containing a quantity of earth. She did so, with little interest at first; but was soon startled to perceive the germ of a plant shooting upward from the soil. Then came the slender stalk; the leaves gradually unfolded themselves; and amid them was a perfect and lovely flower.

"It is magical!" cried Georgiana, "I dare not touch it."

"Nay, pluck it," answered Aylmer,—"pluck it, and inhale its brief perfume while you may. The flower will wither in a few moments and leave nothing save its brown seed vessels; but thence may be perpetuated a race as ephemeral as itself."

But Georgiana had no sooner touched the flower than the whole plant suffered a blight, its leaves turning coal-black as if by the agency of fire.

"There was too powerful a stimulus," said Aylmer, thoughtfully.

To make up for this abortive experiment, he proposed to take her portrait by a scientific process of his own invention. It was to be effected by rays of light striking upon a polished plate of metal. Georgiana assented; but, on looking at the result, was affrighted to find the features of the portrait blurred and indefinable;

while the minute figure of a hand appeared where the cheek should have been. Aylmer snatched the metallic plate and threw it into a jar of corrosive acid.

Soon, however, he forgot these mortifying failures. In the intervals of study and chemical experiment, he came to her flushed and exhausted, but seemed invigorated by her presence, and spoke in glowing language of the resources of his art. He gave a history of the long dynasty of the alchemists, who spent so many ages in quest of the universal solvent, by which the golden principle might be elicited from all things vile and base. Aylmer appeared to believe, that, by the plainest scientific logic, it was altogether within the limits of possibility to discover this long-sought medium; "but," he added, "a philosopher who should go deep enough to acquire the power would attain too lofty a wisdom to stoop to the exercise of it." Not less singular were his opinions in regard to the elixir vitæ. He more than intimated that it was at his option to concoct a liquid that should prolong life for years, perhaps interminably; but that it would produce a discord in nature, which all the world, and chiefly the quaffer of the immortal nostrum, would find cause to curse.

"Aylmer, are you in earnest?" asked Georgiana, looking at him with amazement and fear. "It is terrible to possess such power, or even to dream of possessing it."

"Oh, do not tremble, my love," said her husband, "I would not wrong either you or myself by working such inharmonious effects upon our lives; but I would have you consider how trifling, in comparison, is the skill requisite to remove this little hand."

At the mention of the birthmark, Georgiana, as usual, shrank, as if a redhot iron had touched her cheek.

Again Aylmer applied himself to his labors. She could hear his voice in the distant furnace room giving directions to Aminadab, whose harsh, uncouth, misshapen tones were audible in response, more like the grunt or growl of a brute than human speech. After hours of absence, Aylmer reappeared and proposed that she should now examine his cabinet of chemical products and natural treasures of the earth. Among the former he showed her a small vial, in which, he remarked, was contained a gentle yet most powerful fragrance, capable of impregnating all the breezes that blow across a kingdom. They were of inestimable value, the contents of that little vial; and, as he said so, he threw some of the perfume into the air and filled the room with piercing and invigorating delight.

"And what is this?" asked Georgiana, pointing to a small crystal globe, containing a gold-colored liquid. "It is so beautiful to the eye, that I could imagine it the elixir of life."

"In one sense it is," replied Aylmer; "or rather the elixir of immortality. It is the most precious poison that ever was concocted in this world. By its aid I could apportion the lifetime of any mortal at whom you might point your finger. The strength of the dose would determine whether he were to linger out years, or drop dead in the midst of a breath. No king on his guarded throne could keep his life if I, in my private station, should deem that the welfare of millions justified me in depriving him of it."

"Why do you keep such a terrific drug?" inquired Georgiana in horror.

"Do not mistrust me, dearest," said her husband, smiling; "its virtuous potency is yet greater than its harmful one. But, see! here is a powerful cosmetic. With a few drops of this in a vase of water, freckles may be washed away as easily as the hands are cleansed. A stronger infusion would take the blood out of the cheek, and leave the rosiest beauty a pale ghost."

"Is it with this lotion that you intend to bathe my cheek?" asked Georgiana, anxiously.

"Oh, no," hastily replied her husband; "this is merely superficial. Your case demands a remedy that shall go deeper."

In his interviews with Georgiana, Aylmer generally made minute inquiries as to her sensations, and whether the confinement of the rooms and the temperature of the atmosphere agreed with her. These questions had such a particular drift that Georgiana began to conjecture that she was already subjected to certain physical influences, either breathed in with the fragrant air, or taken with her food. She fancied, likewise, but it might be altogether fancy, that there was a stirring up of her system—a strange, indefinite sensation creeping through her veins, and tingling, half painfully, half pleasurably, at her heart. Still, whenever she dared to look into the mirror, there she beheld herself pale as a white rose and with the crimson birthmark stamped upon her cheek. Not even Aylmer now hated it so much as she.

To dispel the tedium of the hours which her husband found it necessary to devote to the processes of combination and analysis, Georgiana turned over the volumes of his scientific library. In many dark old tomes she met with chapters full of romance and poetry. They were the works of the philosophers of the middle ages, such as Albertus Magnus, Cornelius Agrippa, Paracelsus, and the famous friar who created the prophetic Brazen Head.[2] All these antique naturalists stood in advance of their centuries, yet were imbued with some of their credulity, and therefore were believed, and perhaps imagined themselves to have acquired from the investigation of Nature a power above Nature, and from physics a sway over the spiritual world. Hardly less curious and imaginative were the early volumes of the Transactions of the Royal Society, in which the members, knowing little of the limits of natural possibility, were continually recording wonders, or proposing methods whereby wonders might be wrought.[3]

But to Georgiana the most engrossing volume was a large folio from her husband's own hand, in which he had recorded every experiment of his scientific career, its original aim, the methods adopted for its development, and its final success or failure, with the circumstances to which either event was attributable. The book, in truth, was both the history and emblem of his ardent, ambitious, imaginative, yet practical and laborious life. He handled physical details as if there were nothing beyond them; yet spiritualized them all, and redeemed himself from materialism by his strong and eager aspiration towards the infinite. In his grasp the veriest clod of earth assumed a soul. Georgiana, as she read, reverenced Aylmer and loved him more profoundly than ever, but with a less entire dependence on his judgment than heretofore. Much as he had accomplished, she could not but observe that his most splendid successes were almost invariably failures, if compared with the ideal at which he aimed. His brightest diamonds were the merest pebbles, and felt to be so by himself, in comparison with the inestimable gems which lay hidden beyond his reach. The volume, rich with achievements that had won renown for its author, was yet as melancholy a record as ever mortal hand had penned. It was the sad confession, and continual exemplification, of the shortcomings of the composite man, the spirit burthened with clay and working in matter, and of the despair that assails the higher nature at finding itself so miserably thwarted by the earthly part. Perhaps every man of genius in whatever sphere might recognize the image of his own experience in Aylmer's journal.

So deeply did these reflections affect Georgiana that she laid her face upon the open volume and burst into tears. In this situation she was found by her husband.

"It is dangerous to read in a sorcerer's books," said he with a smile, though his countenance was uneasy and displeased. "Georgiana, there are pages in that volume which I can scarcely glance over and keep my senses. Take heed lest it prove as detrimental to you."

"It has made me worship you more than ever," said she.

"Ah, wait for this one success," rejoined he, "then worship me if you will. I shall deem myself hardly unworthy of it. But come, I have sought you for the luxury of your voice. Sing to me, dearest!"

So she poured out the liquid music of her voice to quench the thirst of his spirit. He then took his leave with a boyish exuberance of gayety, assuring her that her seclusion would endure but a little longer, and that the result was already certain. Scarcely had he departed when Georgiana felt irresistibly impelled to follow him. She had forgotten to inform Aylmer of a symptom which for two or three hours past had begun to excite her attention. It was a sensation in the fatal birthmark, not painful, but which induced a restlessness throughout her system. Hastening after her husband, she intruded for the first time into the laboratory.

The first thing that struck her eye was the furnace, that hot and feverish worker, with the intense glow of its fire, which by the quantities of soot clustered above it seemed to have been burning for ages. There was a distilling apparatus in full operation. Around the room were retorts, tubes, cylinders, crucibles, and other apparatus of chemical research. An electrical machine stood ready for immediate use. The atmosphere felt oppressively close, and was tainted with gaseous odors, which had been tormented forth by the processes of science. The severe and homely simplicity of the apartment, with its naked walls and brick pavement, looked strange, accustomed as Georgiana had become to the fantastic elegance of her boudoir. But what chiefly, indeed almost solely, drew her attention, was the aspect of Aylmer himself.

He was pale as death, anxious and absorbed, and hung over the furnace as if it depended upon his utmost watchfulness whether the liquid which it was distilling should be the draught of immortal happiness or misery. How different from the sanguine and joyous mien that he had assumed for Georgiana's encouragement!

"Carefully now, Aminadab; carefully, thou human machine; carefully, thou man of clay!" muttered Aylmer, more to himself than his assistant. "Now, if there be a thought too much or too little, it is all over."

"Ho! ho!" mumbled Aminadab. "Look, master! look!"

Aylmer raised his eyes hastily, and at first reddened, then grew paler than ever, on beholding Georgiana. He rushed towards her and seized her arm with a gripe that left the print of his fingers upon it.

"Why do you come hither? Have you no trust in your husband?" cried he, impetuously. "Would you throw the blight of that fatal birthmark over my labors? It is not well done. Go, prying woman, go!"

"Nay, Aylmer," said Georgiana, with the firmness of which she possessed no stinted endowment, "it is not you that have a right to complain. You mistrust your wife; you have concealed the anxiety with which you watch the development of this experiment. Think not so unworthily of me, my husband. Tell me all the risk we run, and fear not that I shall shrink; for my share in it is far less than your own."

"No, no, Georgiana!" said Aylmer impatiently; "it must not be."

"I submit," replied she calmly. "And, Aylmer, I shall quaff whatever draught you bring me; but it will be on the same principle that would induce me to take a dose of poison, if offered by your hand."

"My noble wife," said Aylmer, deeply moved, "I knew not the height and depth of your nature, until now. Nothing shall be concealed. Know, then, that this crimson hand, superficial as it seems, has clutched its grasp into your being with a strength of which I had no previous conception. I have already administered agents powerful enough to do aught except to change your entire physical system. Only one thing remains to be tried. If that fail us we are ruined."

"Why did you hesitate to tell me this?" asked she.

"Because, Georgiana," said Aylmer, in a low voice, "there is danger."

"Danger? There is but one danger—that this horrible stigma shall be left upon my cheek!" cried Georgiana. "Remove it, remove it, whatever be the cost, or we shall both go mad!"

"Heaven knows, your words are too true," said Aylmer, sadly. "And now, dearest, return to your boudoir. In a little while all will be tested."

He conducted her back and took leave of her with a solemn tenderness, which spoke far more than his words how much was now at stake. After his departure Georgiana became rapt in musings. She considered the character of Aylmer, and did it completer justice than at any previous moment. Her heart exulted, while it trembled, at his honorable love—so pure and lofty that it would accept nothing less than perfection nor miserably make itself contented with an earthlier nature than he had dreamed of. She felt how much more precious was such a sentiment than that meaner kind which would have borne with the imperfection for her sake, and have been guilty of treason to holy love by degrading its perfect idea to the level of the actual. And, with her whole spirit, she prayed that, for a single moment, she might satisfy his highest and deepest conception. Longer than one moment, she well knew, it could not be; for his spirit was ever on the march, ever ascending, and each instant required something that was beyond the scope of the instant before.

The sound of her husband's footsteps aroused her. He bore a crystal goblet, containing a liquor colorless as water, but bright enough to be the draught of immortality. Aylmer was pale; but it seemed rather the consequence of a highly wrought state of mind and tension of spirit than of fear or doubt.

"The concoction of the draught has been perfect," said he, in answer to Georgiana's look. "Unless all my science have deceived me, it cannot fail."

"Save on your account, my dearest Aylmer," observed his wife, "I might wish to put off this birthmark of mortality by relinquishing mortality itself in preference to any other mode. Life is but a sad possession to those who have attained precisely the degree of moral advancement at which I stand. Were I weaker and blinder it might be happiness. Were I stronger, it might be endured hopefully. But, being what I find myself, methinks I am of all mortals the most fit to die."

"You are fit for heaven without tasting death!" replied her husband. "But why do we speak of dying? The draught cannot fail. Behold its effect upon this plant."

On the window seat there stood a geranium diseased with yellow blotches, which had overspread all its leaves. Aylmer poured a small quantity of the liquid upon the soil in which it grew. In a little time, when the roots of the plant had taken up the moisture, the unsightly blotches began to be extinguished in a living verdure.

"There needed no proof," said Georgiana, quietly. "Give me the goblet. I joyfully stake all upon your word."

"Drink, then, thou lofty creature!" exclaimed Aylmer, with fervid admiration. "There is no taint of imperfection on thy spirit. Thy sensible frame, too, shall soon be all perfect."

She quaffed the liquid and returned the goblet to his hand.

"It is grateful," said she, with a placid smile. "Methinks it is like water from a heavenly fountain; for it contains I know not what of unobtrusive fragrance and deliciousness. It allays a feverish thirst that had parched me for many days. Now, dearest, let me sleep. My earthly senses are closing over my spirit like the leaves around the heart of a rose at sunset."

She spoke the last words with a gentle reluctance, as if it required almost more energy than she could command to pronounce the faint and lingering syllables. Scarcely had they loitered through her lips ere she was lost in slumber. Aylmer sat by her side, watching her aspect with the emotions proper to a man the whole value of whose existence was involved in the process now to be tested. Mingled with this mood, however, was the philosophic investigation characteristic of the man of science. Not the minutest symptom escaped him. A heightened flush of the cheek, a slight irregularity of breath, a quiver of the eyelid, a hardly perceptible tremor through the frame,—such were the details which, as the moments passed, he wrote down in his folio volume. Intense thought had set its stamp upon every previous page of that volume, but the thoughts of years were all concentrated upon the last.

While thus employed, he failed not to gaze often at the fatal hand, and not without a shudder. Yet once, by a strange and unaccountable impulse, he pressed it with his lips. His spirit recoiled, however, in the very act; and Georgiana, out of the midst of her deep sleep, moved uneasily and murmured, as if in remonstrance. Again Aylmer resumed his watch. Nor was it without avail. The crimson hand, which at first had been strongly visible upon the marble paleness of Georgiana's cheek, now grew more faintly outlined. She remained not less pale than ever; but the birthmark, with every breath that came and went, lost somewhat of its former distinctness. Its presence had been awful; its departure was more awful still. Watch the stain of the rainbow fading out of the sky, and you will know how that mysterious symbol passed away.

"By Heaven, it is well nigh gone!" said Aylmer to himself, in almost irrepressible ecstasy. "I can scarcely trace it now. Success! Success! And now it is like the faintest rose color. The slightest flush of blood across her cheek would overcome it. But she is so pale!"

He drew aside the window curtain, and suffered the light of natural day to fall into the room and rest upon her cheek. At the same time, he heard a gross, hoarse chuckle, which he had long known as his servant Aminadab's expression of delight.

"Ah, clod! Ah, earthly mass!" cried Aylmer, laughing in a sort of frenzy, "you have served me well! Matter and spirit—earth and heaven—have both done their part in this! Laugh, thing of the senses! You have earned the right to laugh."

These exclamations broke Georgiana's sleep. She slowly unclosed her eyes and gazed into the mirror which her husband had arranged for that purpose. A faint smile flitted over her lips, when she recognized how barely perceptible was now that crimson hand which had once blazed forth with such disastrous brilliancy as to scare away all their happiness. But then her eyes sought Aylmer's face with a trouble and anxiety that he could by no means account for.

"My poor Aylmer!" murmured she.

"Poor? Nay, richest, happiest, most favored!" exclaimed he. "My peerless bride, it is successful! You are perfect!"

"My poor Aylmer," she repeated, with a more than human tenderness, "you have aimed loftily; you have done nobly. Do not repent that with so high and pure a feeling, you have rejected the best the earth could offer. Aylmer, dearest Aylmer, I am dying!"

Alas, it was too true! The fatal hand had grappled with the mystery of life, and was the bond by which an angelic spirit kept itself in union with a mortal frame. As the last crimson tint of the birthmark—that sole token of human imperfection—faded from her cheek, the parting breath of the now perfect woman passed into the atmosphere, and her soul, lingering a moment near her husband, took its heavenward flight. Then a hoarse, chuckling laugh was heard again! Thus ever does the gross fatality of earth exult in its invariable triumph over the immortal essence which, in this dim sphere of half-development, demands the completeness of a higher state. Yet, had Aylmer reached a profounder wisdom, he need not thus have flung away the happiness which would have woven his mortal life of the selfsame texture with the celestial. The momentary circumstance was too strong for him; he failed to look beyond the shadowy scope of time, and, living once for all in eternity, to find the perfect future in the present.

DISCUSSION QUESTIONS

1. What motivates Aylmer as a doctor-scientist? What role does the history of alchemy play in the type of medicine he practices? Has he been successful in his career prior to his marriage to Georgiana? Does it matter that Aylmer and Georgiana are newlyweds?

2. Why is it significant that Georgiana's birthmark appears different to various beholders? Compare the various responses to her birthmark and what each indicates about the viewer's motivations. How is the act of "seeing" our own and others' bodies shaped by personal, societal, and cultural norms? Are health care providers trained to interpret what they see with these norms in mind—or is a birthmark, for example, always just a physical sign that holds the same meaning in every context?

3. What role do appearances play in our perceptions of health and illness today? Are those whose illnesses are visible to the public (traumatized or disabled bodies, for example) perceived or treated differently than those whose illnesses are hidden from sight? In the context of health care, do personal biases and assumptions regarding appearances influence the way individual providers treat patients? How important is it for a health care provider to be aware of his or her own biases? Do any systemic biases related to appearance exist in health care?

4. Aylmer develops a technology that he believes will heal his wife but that ultimately destroys her. List some of the risks associated with twenty-first century advances in medical, pharmaceutical, or scientific technologies. Who decides what level of risk is acceptable? In what ways do technologies affect the patient-provider relationship negatively? How might patients and providers guard against these negative effects?

5. Women in the nineteenth century, particularly lower classes of women and slave women, were vulnerable to experimental health care practices. What qualities do we associate with vulnerable populations today, and how might the health humanities help providers and researchers address issues associated with the health care disparities and inequalities these populations often face?

Egotism;* or, The Bosom Serpent

From the unpublished "Allegories of the Heart"

March 1843, *United States Magazine and Democratic Review*

Could a snake or an electric eel invade and take up residence in a human body? Many nineteenth-century Americans believed so, thanks to a rich and varied tradition of bosom serpent literature, lore, and medical case reports. Before undertaking his own version of the serpent tale, Hawthorne had likely seen a number of literary and popular references to the motif; they came to him from four different identifiable sources: Spenser's The Faerie Queen, *historical medical cases, nineteenth-century newspaper reports, and theological writings. A story published in the February 25, 1837, edition of the* Philadelphia Public Ledger *as "Extraordinary Case" was very likely the most direct source for "Egotism" (Bush 1971).[1]*

Hawthorne, of course, transforms the tale into an allegory that offers universal and timeless themes related to human illness and suffering, while also presenting a glimpse into the evolving theories and practices surrounding mental illness in America. Almost a century before "Egotism; or, The Bosom Serpent" was published, physician Benjamin Rush treated mentally ill patients with bloodletting, believing hypertension in the brain—not, as religious theorists argued, spiritual turpitude or demonic possession—to be the cause of their suffering. In general, physicians believed all mental illness to be associated with brain trauma, and doctors were eager to study the brains of mental patients to identify lesions responsible for derangement. The first hospitals established exclusively for the treatment of individuals with mental illness in the late 1700s followed the model of French psychiatrist Philippe Pinel, promoting the "moral treatment," an individualized approach that involved not only physical therapies like bloodletting, cold baths, and doses of morphine but also medicine for the spirit that included training in religion and the arts (Goodwin 1999). By the mid-nineteenth century, practitioners of the new clinical medicine focused on the body as the source of all illness, but religious theories persisted as well (Long 1989). Women's mental illnesses were often, incorrectly, linked to female reproductive organs, such as the uterus (Theriot 1990; Studd 2007).

---

*The physical fact, to which it is here attempted to give a moral signification, has been known to occur in more than one instance.

In short, treatment of individuals with mental illness was still plagued by great uncertainty. Moreover, as the American population grew, the numbers of those suffering from mental illness grew as well, and asylums that had originally been designed as humane institutions, where patients went to heal, became prisons for those the community deemed unfit for society. As numbers of patients rose, individualized care disappeared, and the level of healing in asylums declined (Benjamin and Baker 2004). For patients like Hawthorne's Roderick Elliston, whom doctors ultimately deem sane, there was no healing at all—at least not from any medical practitioner.

"Egotism" gives us a glimpse of available mental health treatments around mid-century, some forty years before "new psychology," or "experimental psychology," a name meant to stress the new psychology's scientific basis. While Hawthorne does take us into the clinic in this tale, he does so before there was such a thing as clinical psychology or, in the community, counseling psychology (Benjamin and Baker 2004). The tale calls us to consider the methods of healing available to Elliston from the medical community and to examine his attempts at self-healing. A close examination of the latter should provoke discussions of the role of storytelling and the arts in health and healing.

•

"Here he comes!" shouted the boys along the street. "Here comes the man with a snake in his bosom!"

This outcry, saluting Herkimer's ears as he was about to enter the iron gate of the Elliston mansion, made him pause. It was not without a shudder that he found himself on the point of meeting his former acquaintance, whom he had known in the glory of youth, and whom now after an interval of five years, he was to find the victim either of a diseased fancy or a horrible physical misfortune.

"A snake in his bosom!" repeated the young sculptor to himself. "It must be he. No second man on earth has such a bosom friend. And now, my poor Rosina, Heaven grant me wisdom to discharge my errand aright! Woman's faith must be strong indeed since thine has not yet failed."

Thus musing, he took his stand at the entrance of the gate and waited until the personage so singularly announced should make his appearance. After an instant or two he beheld the figure of a lean man, of unwholesome look, with glittering eyes and long black hair, who seemed to imitate the motion of a snake; for, instead of walking straight forward with open front, he undulated along the pavement in a curved line. It may be too fanciful to say that something, either in his moral or material aspect, suggested the idea that a miracle had been wrought by transforming a serpent into a man, but so imperfectly that the snaky nature was yet hidden, and scarcely hidden, under the mere outward guise of humanity. Herkimer remarked that his complexion had a greenish tinge over its sickly white, reminding him of a species of marble out of which he had once wrought a head of Envy, with her snaky locks.

The wretched being approached the gate, but, instead of entering, stopped short and fixed the glitter of his eye full upon the compassionate yet steady countenance of the sculptor.

"It gnaws me! It gnaws me!" he exclaimed.

And then there was an audible hiss, but whether it came from the apparent lunatic's own lips, or was the real hiss of a serpent, might admit of discussion. At all events, it made Herkimer shudder to his heart's core.

"Do you know me, George Herkimer?" asked the snake-possessed.

Herkimer did know him; but it demanded all the intimate and practical acquaintance with the human face, acquired by modelling actual likenesses in clay, to recognize the features of Roderick Elliston in the visage that now met the sculptor's gaze. Yet it was he. It added nothing to the wonder to reflect that the once brilliant young man had undergone this odious and fearful change during the no more than five brief years of Herkimer's abode at Florence. The possibility of such a transformation being granted, it was as easy to conceive it effected in a moment as in an age. Inexpressibly shocked and startled, it was still the keenest pang when Herkimer remembered that the fate of his cousin Rosina, the ideal of gentle womanhood, was indissolubly interwoven with that of a being whom Providence seemed to have unhumanized.

"Elliston! Roderick!" cried he, "I had heard of this; but my conception came far short of the truth. What has befallen you? Why do I find you thus?"

"Oh, 'tis a mere nothing! A snake! A snake! The commonest thing in the world. A snake in the bosom—that's all," answered Roderick Elliston. "But how is your own breast?" continued he, looking the sculptor in the eye with the most acute and penetrating glance that it had ever been his fortune to encounter. "All pure and wholesome? No reptile there? By my faith and conscience, and by the devil within me, here is a wonder! A man without a serpent in his bosom!"

"Be calm, Elliston," whispered George Herkimer, laying his hand upon the shoulder of the snake-possessed. "I have crossed the ocean to meet you. Listen! Let us be private. I bring a message from Rosina—from your wife!"

"It gnaws me! It gnaws me!" muttered Roderick.

With this exclamation, the most frequent in his mouth, the unfortunate man clutched both hands upon his breast as if an intolerable sting or torture impelled him to rend it open, and let out the living mischief, even where it intertwined with his own life. He then freed himself from Herkimer's grasp by a subtle motion, and, gliding through the gate, took refuge in his antiquated family residence. The sculptor did not pursue him. He saw that no available intercourse could be expected at such a moment, and was desirous, before another meeting, to inquire closely into the nature of Roderick's disease and the circumstances that had reduced him to so lamentable a condition. He succeeded in obtaining the necessary information from an eminent medical gentleman.

Shortly after Elliston's separation from his wife—now nearly four years ago—his associates had observed a singular gloom spreading over his daily life, like those chill, gray mists that sometimes steal away the sunshine from a summer's morning. The symptoms caused them endless perplexity. They knew not whether ill health were robbing his spirits of elasticity, or whether a canker of the mind was gradually eating, as such cankers do, from his moral system into the physical frame, which is but the shadow of the former. They looked for the root of this trouble in his shattered schemes of domestic bliss,—wilfully shattered by himself,—but could not be satisfied of its existence there. Some thought that their once brilliant friend was in an incipient stage of insanity, of which his passionate impulses had perhaps been the forerunners; others prognosticated a general blight and gradual decline. From Roderick's own lips, they could learn nothing. More than once, it is true, he had been heard to say, clutching his hands convulsively upon his breast—"It gnaws me! It gnaws me!"—but, by different auditors, a great diversity of explanation was assigned to this ominous expression. What could it be that gnawed the breast of Roderick Elliston? Was it sorrow? Was it merely the tooth of physical disease? Or, in his reckless course, often verging upon profligacy, if not plunging into its depths, had he been guilty of some deed, which made his bosom a prey to the deadlier fangs of remorse? There was plausible ground for each of these conjectures; but it must not be concealed that more than one elderly gentleman, the victim of good cheer and slothful habits, magisterially pronounced the secret of the whole matter to be Dyspepsia!

Meanwhile, Roderick seemed aware how generally he had become the subject of curiosity and conjecture, and, with a morbid repugnance to such notice, or to any notice whatsoever, estranged himself from all companionship. Not merely the eye of man was a horror to him; not merely the light of a friend's countenance; but even the blessed sunshine, likewise, which in its universal beneficence typifies the radiance of the Creator's face, expressing his love for all the creatures of his hand. The dusky twilight was now too transparent for Roderick Elliston; the blackest midnight was his chosen hour to steal abroad; and if ever he were seen, it was when the watchman's lantern gleamed upon his figure, gliding along the street with his hands clutched upon his bosom, still muttering, "It gnaws me! It gnaws me!" What could it be that gnawed him?

After a time, it became known that Elliston was in the habit of resorting to all the noted quacks that infested the city, or whom money would tempt to journey thither from a distance. By one of these persons, in the exultation of a supposed cure, it was proclaimed far and wide, by dint of handbills and little pamphlets on dingy paper, that a distinguished gentleman, Roderick Elliston, Esq., had been relieved of a SNAKE in his stomach! So here was the monstrous secret, ejected from its lurking place into public view, in all its horrible deformity. The mystery was out, but not so the bosom serpent. He, if it were anything but a delusion, still lay coiled in his living den. The empiric's cure had been a sham, the effect, it was supposed, of some stupefying drug which more nearly caused the death of the patient than of the odious reptile that possessed him. When Roderick Elliston regained entire sensibility, it was to find his misfortune the town talk—the more than nine days' wonder and horror—while, at his bosom, he felt the sickening motion of a thing alive, and the gnawing of that restless fang which seemed to gratify at once a physical appetite and a fiendish spite.

He summoned the old black servant, who had been bred up in his father's house, and was a middle-aged man while Roderick lay in his cradle.

"Scipio!" he began; and then paused, with his arms folded over his heart. "What do people say of me, Scipio."

"Sir! my poor master! that you had a serpent in your bosom," answered the servant with hesitation.

"And what else?" asked Roderick, with a ghastly look at the man.

"Nothing else, dear master," replied Scipio, "only that the doctor gave you a powder, and that the snake leapt out upon the floor."

"No, no!" muttered Roderick to himself, as he shook his head, and pressed his hands with a more convulsive force upon his breast, "I feel him still. It gnaws me! It gnaws me!"

From this time, the miserable sufferer ceased to shun the world, but rather solicited and forced himself upon the notice of acquaintances and strangers. It was partly the result of desperation on finding that the cavern of his own bosom had not proved deep and dark enough to hide the secret, even while it was so secure a fortress for the loathsome fiend that had crept into it. But still more, this craving for notoriety was a symptom of the intense morbidness which now pervaded his nature. All persons chronically diseased are egotists, whether the disease be of the mind or body; whether sin, sorrow, or merely the more tolerable calamity of some endless pain, or mischief among the cords of mortal life. Such individuals are made acutely conscious of a self, by the torture in which it dwells. Self, therefore, grows to be so prominent an object with them that they cannot but present it to the face of every casual passer-by. There is a pleasure—perhaps the greatest of which the sufferer is susceptible—in display-ing the wasted or ulcerated limb, or the cancer in the breast; and the fouler the crime, with so much the more difficulty does the perpetrator prevent it from thrusting up its snake-like head to frighten the world; for it is that cancer, or that crime, which constitutes their respective individuality. Roderick Elliston, who, a little while before, had held himself so scornfully above the common lot of men, now paid full allegiance to this humiliating law. The snake in his bosom seemed the symbol of a monstrous egotism, to which everything was referred, and which he pampered, night and day, with a continual and exclusive sacrifice of devil worship.

He soon exhibited what most people considered indubitable tokens of in-sanity. In some of his moods, strange to say, he prided and gloried himself on being marked out from the ordinary experience of mankind, by the possession of a double nature, and a life within a life. He appeared to imagine that the snake was a divinity,—not celestial, it is true, but darkly infernal,—and that he thence derived an eminence and a sanctity, horrid, indeed, yet more desirable than whatever ambition aims at. Thus he drew his misery around him like a regal mantle, and looked down triumphantly upon those whose vitals nourished no deadly monster. Oftener, however, his human nature asserted its empire over him in the shape of a yearning for fellowship. It grew to be his custom to spend the whole day in wandering about the streets, aimlessly, unless it might be called an aim to establish a species of brotherhood between himself and the world. With cankered ingenuity, he sought out his own disease in every breast. Whether insane or not, he showed so keen a perception of frailty, error, and vice, that many persons gave him credit for being possessed not merely with a serpent, but with an actual fiend, who imparted this evil faculty of recognizing whatever was ugliest in man's heart.

For instance, he met an individual, who, for thirty years, had cherished a hatred against his own brother. Roderick, amidst the throng of the street, laid his hand on this man's chest, and looking full into his forbidding face,—

"How is the snake to-day?" he inquired, with a mock expression of sympathy.

"The snake!" exclaimed the brother hater—"what do you mean?"

"The snake! The snake! Does he gnaw you?" persisted Roderick. "Did you take counsel with him this morning when you should have been saying your prayers? Did he sting, when you thought of your brother's health, wealth, and good repute? Did he caper for joy, when you remembered the profligacy of his only son? And whether he stung, or whether he frolicked, did you feel his poison throughout your body and soul, converting everything to sourness and bitterness? That is the way of such serpents. I have learned the whole nature of them from my own!"

"Where is the police?" roared the object of Roderick's persecution, at the same time giving an instinctive clutch to his breast. "Why is this lunatic allowed to go at large?"

"Ha, ha!" chuckled Roderick, releasing his grasp of the man. "His bosom serpent has stung him then!"

Often it pleased the unfortunate young man to vex people with a lighter satire, yet still characterized by somewhat of snakelike virulence. One day he encountered an ambitious statesman, and gravely inquired after the welfare of his boa constrictor; for of that species, Roderick affirmed, this gentleman's serpent must needs be, since its appetite was enormous enough to devour the whole country and constitution. At another time, he stopped a close-fisted old fellow, of great wealth, but who skulked about the city in the guise of a scarecrow, with a patched blue surtout, brown hat, and mouldy boots, scraping pence together, and picking up rusty nails. Pretending to look earnestly at this respectable person's stomach, Roderick assured him that his snake was a copper-head, and had been generated by the immense quantities of that base metal, with which he daily defiled his fingers. Again, he assaulted a man of rubicund visage, and told him that few bosom serpents had more of the devil in them than those that breed in the vats of a distillery. The next whom Roderick honored with his attention was a distinguished clergyman, who happened just then to be engaged in a theological controversy, where human wrath was more perceptible than divine inspiration.

"You have swallowed a snake in a cup of sacramental wine," quoth he.

"Profane wretch!" exclaimed the divine; but, nevertheless, his hand stole to his breast.

He met a person of sickly sensibility, who, on some early disappointment, had retired from the world, and thereafter held no intercourse with his fellow-men, but brooded sullenly or passionately over the irrevocable past. This man's very heart, if Roderick might be believed, had been changed into a serpent, which would finally torment both him and itself to death. Observing a married couple, whose domestic troubles were matter of notoriety, he condoled with both on having mutually taken a house adder to their bosoms. To an envious author, who depreciated works which he could never equal, he said that his snake was the slimiest and filthiest of all the reptile tribe, but was fortunately without a sting. A man of impure life, and a brazen face, asking Roderick if there were any serpent in his breast, he told him that there was, and of the same species that once tortured Don Rodrigo, the Goth. He took a fair young girl by the hand, and gazing sadly into her eyes, warned her that she cherished a serpent of the deadliest kind within her gentle breast; and the world found the truth of those ominous words, when, a few months afterwards, the poor girl died of love and shame. Two ladies, rivals in fashionable life, who tormented one another with a thousand little stings of womanish spite, were given to understand that each of their hearts was a nest of diminutive snakes, which did quite as much mischief as one great one.

But nothing seemed to please Roderick better than to lay hold of a person infected with jealousy, which he represented as an enormous green reptile, with an ice-cold length of body, and the sharpest sting of any snake save one.

"And what one is that?" asked a by-stander, overhearing him.

It was a dark-browed man, who put the question; he had an evasive eye, which in the course of a dozen years had looked no mortal directly in the face. There was an ambiguity about this person's character,—a stain upon his repu-tation,—yet none could tell precisely of what nature, although the city gossips, male and female, whispered the most atrocious surmises. Until a recent period he had followed the sea, and was, in fact, the very shipmaster whom George Herkimer had encountered, under such singular circumstances, in the Grecian Archipelago.

"What bosom serpent has the sharpest sting?" repeated this man; but he put the question as if by a reluctant necessity, and grew pale while he was uttering it.

"Why need you ask?" replied Roderick, with a look of dark intelligence. "Look into your own breast! Hark! my serpent bestirs himself! He acknowledges the presence of a master fiend!"

And then, as the by-standers afterwards affirmed, a hissing sound was heard, apparently in Roderick Elliston's breast. It was said, too, that an answering hiss came from the vitals of the shipmaster, as if a snake were actually lurking there and had been aroused by the call of its brother reptile. If there were in fact any such sound, it might have been caused by a malicious exercise of ventriloquism on the part of Roderick.

Thus making his own actual serpent—if a serpent there actually was in his bosom—the type of each man's fatal error, or hoarded sin, or unquiet con-science, and striking his sting so unremorsefully into the sorest spot, we may well imagine that Roderick became the pest of the city. Nobody could elude him—none could withstand him. He grappled with the ugliest truth that he could lay his hand on, and compelled his adversary to do the same. Strange spectacle in human life where it is the instinctive effort of one and all to hide those sad realities, and leave them undisturbed beneath a heap of superficial topics which constitute the materials of intercourse between man and man! It was not to be tolerated that Roderick Elliston should break through the tacit compact by which the world has done its best to secure repose without relinquishing evil. The victims of his malicious remarks, it is true, had brothers enough to keep them in countenance; for, by Roderick's theory, every mortal bosom harbored either a brood of small serpents, or one overgrown monster that had devoured all the rest. Still the city could not bear this new apostle. It was demanded by nearly all, and particularly by the most respectable inhabitants, that Roderick should no longer be permitted to violate the received rules of decorum by obtruding his own bosom serpent to the public gaze, and dragging those of decent people from their lurking places.

Accordingly, his relatives interfered and placed him in a private asylum for the insane. When the news was noised abroad, it was observed that many persons walked the streets with freer countenances and covered their breasts less carefully with their hands.

His confinement, however, although it contributed not a little to the peace of the town, operated unfavorably upon Roderick himself. In solitude his melancholy grew more black and sullen. He spent whole days—indeed, it was his sole occupation—in communing with the serpent. A conversation was sustained, in which, as it seemed, the hidden monster bore a part, though unintelligibly to the listeners, and inaudible except in a hiss. Singular as it may appear, the sufferer had now contracted a sort of affection for his tormentor, mingled, however, with the intensest loathing and horror. Nor were such discordant emotions incompatible. Each, on the contrary, imparted strength and poignancy to its opposite. Horrible love—horrible antipathy—embracing one another in his bosom, and both concentrating themselves upon a being that had crept into his vitals, or been engendered there, and which was nourished with his food, and lived upon his life, and was as intimate with him as his own heart, and yet was the foulest of all created things! But not the less was it the true type of a morbid nature.

Sometimes, in his moments of rage and bitter hatred against the snake and himself, Roderick determined to be the death of him, even at the expense of his own life. Once he attempted it by starvation; but, while the wretched man was on the point of famishing, the monster seemed to feed upon his heart, and to thrive and wax gamesome, as if it were his sweetest and most congenial diet. Then he privily took a dose of active poison, imagining that it would not fail to kill either himself, or the devil that possessed him, or both together. Another mistake; for if Roderick had not yet been destroyed by his own poisoned heart nor the snake by gnawing it, they had little to fear from arsenic or corrosive sublimate. Indeed, the venomous pest appeared to operate as an antidote against all other poisons. The physicians tried to suffocate the fiend with tobacco smoke. He breathed it as freely as if it were his native atmosphere. Again, they drugged their patient with opium and drenched him with intoxicating liquors, hoping that the snake might thus be reduced to stupor and perhaps be ejected from the stomach. They succeeded in rendering Roderick insensible; but, placing their hands upon his breast, they were inexpressibly horror stricken to feel the monster wriggling, twining, and darting to and fro within his narrow limits, evidently enlivened by the opium or alcohol, and incited to unusual feats of activity. Thenceforth they gave up all attempts at cure or palliation. The doomed sufferer submitted to his fate, resumed his former loathsome affection for the bosom fiend, and spent whole miserable days before a looking-glass, with his mouth wide open, watching, in hope and horror, to catch a glimpse of the snake's head, far down within his throat. It is supposed that he succeeded; for the attendants once heard a frenzied shout, and, rushing into the room, found Roderick lifeless upon the floor.

He was kept but little longer under restraint. After minute investigation, the medical directors of the asylum decided that his mental disease did not amount to insanity, nor would warrant his confinement, especially as its influence upon his spirits was unfavorable, and might produce the evil which it was meant to remedy. His eccentricities were doubtless great; he had habitually violated many of the customs and prejudices of society; but the world was not, without surer ground, entitled to treat him as a madman. On this decision of such competent authority, Roderick was released, and had returned to his native city the very day before his encounter with George Herkimer.

As soon as possible after learning these particulars the sculptor, together with a sad and tremulous companion, sought Elliston at his own house. It was a large, sombre edifice of wood, with pilasters and a balcony, and was divided from one of the principal streets by a terrace of three elevations, which was ascended by successive flights of stone steps. Some immense old elms almost concealed the front of the mansion. This spacious and once magnificent family residence was built by a grandee of the race, early in the past century, at which epoch, land being of small comparative value, the garden and other grounds had formed quite an extensive domain. Although a portion of the ancestral heritage had been alienated, there was still a shadowy enclosure in the rear of the mansion where a student, or a dreamer, or a man of stricken heart might lie all day upon the grass, amid the solitude of murmuring boughs, and forget that a city had grown up around him.

Into this retirement, the sculptor and his companion were ushered by Scipio, the old black servant, whose wrinkled visage grew almost sunny with intelligence and joy as he paid his humble greetings to one of the two visitors.

"Remain in the arbor," whispered the sculptor to the figure that leaned upon his arm. "You will know whether, and when, to make your appearance."

"God will teach me," was the reply. "May he support me too!"

Roderick was reclining on the margin of a fountain which gushed into the fleckered sunshine with the same clear sparkle and the same voice of airy quietude as when trees of primeval growth flung their shadows across its bosom. How strange is the life of a fountain!—born at every moment, yet of an age coeval with the rocks, and far surpassing the venerable antiquity of a forest.

"You are come! I have expected you," said Elliston, when he became aware of the sculptor's presence.

His manner was very different from that of the preceding day—quiet, courteous, and, as Herkimer thought, watchful both over his guest and himself. This unnatural restraint was almost the only trait that betokened anything amiss. He had just thrown a book upon the grass, where it lay half opened, thus disclosing itself to be a natural history of the serpent tribe, illustrated by lifelike plates. Near it lay that bulky volume, the Ductor Dubitantium of Jeremy Taylor, full of cases of conscience, and in which most men, possessed of a conscience, may find something applicable to their purpose.

"You see," observed Elliston, pointing to the book of serpents, while a smile gleamed upon his lips, "I am making an effort to become better acquainted with my bosom friend; but I find nothing satisfactory in this volume. If I mistake not, he will prove to be *sui generis,* and akin to no other reptile in creation."

"Whence came this strange calamity?" inquired the sculptor.

"My sable friend Scipio has a story," replied Roderick, "of a snake that had lurked in this fountain—pure and innocent as it looks—ever since it was known to the first settlers. This insinuating personage once crept into the vitals of my great grandfather and dwelt there many years, tormenting the old gentleman beyond mortal endurance. In short it is a family peculiarity. But, to tell you the truth, I have no faith in this idea of the snake's being an heirloom. He is my own snake, and no man's else."

"But what was his origin?" demanded Herkimer.

"Oh, there is poisonous stuff in any man's heart, sufficient to generate a brood of serpents," said Elliston, with a hollow laugh. "You should have heard my homilies to the good town's-people. Positively, I deem myself fortunate in having bred but a single serpent. You, however, have none in your bosom, and therefore cannot sympathize with the rest of the world. It gnaws me! It gnaws me!"

With this exclamation, Roderick lost his self-control and threw himself upon the grass, testifying his agony by intricate writhings, in which Herkimer could not but fancy a resemblance to the motions of a snake. Then, likewise, was heard that frightful hiss, which often ran through the sufferer's speech, and crept between the words and syllables, without interrupting their succession.

"This is awful indeed!" exclaimed the sculptor—"an awful infliction, whether it be actual or imaginary! Tell me, Roderick Elliston, is there any remedy for this loathsome evil?"

"Yes, but an impossible one," muttered Roderick, as he lay wallowing with his face in the grass. "Could I for one instant forget myself, the serpent might not abide within me. It is my diseased self-contemplation that has engendered and nourished him."

"Then forget yourself, my husband," said a gentle voice above him; "forget yourself in the idea of another!"

Rosina had emerged from the arbor, and was bending over him, with the shadow of his anguish reflected in her countenance, yet so mingled with hope and unselfish love that all anguish seemed but an earthly shadow and a dream. She touched Roderick with her hand. A tremor shivered through his frame. At that moment, if report be trustworthy, the sculptor beheld a waving motion through the grass, and heard a tinkling sound, as if something had plunged into the fountain. Be the truth as it might, it is certain that Roderick Elliston sat up like a man renewed, restored to his right mind, and rescued from the fiend which had so miserably overcome him in the battle-field of his own breast.

"Rosina!" cried he, in broken and passionate tones, but with nothing of the wild wail that had haunted his voice so long, "forgive! forgive!"

Her happy tears bedewed his face.

"The punishment has been severe," observed the sculptor. "Even Justice might now forgive; how much more a woman's tenderness! Roderick Elliston, whether the serpent was a physical reptile, or whether the morbidness of your nature suggested that symbol to your fancy, the moral of the story is not the less true and strong. A tremendous Egotism, manifesting itself in your case in the form of jealousy, is as fearful a fiend as ever stole into the human heart. Can a breast, where it has dwelt so long, be purified?"

"Oh, yes," said Rosina with a heavenly smile. "The serpent was but a dark fantasy, and what it typified was as shadowy as itself. The past, dismal as it seems, shall fling no gloom upon the future. To give it its due importance we must think of it but as an anecdote in our Eternity."

DISCUSSION QUESTIONS

1. A number of Roderick's associates have engaged in conjecture about the nature of his illness. Reread the passage about the various theories these associates have developed: what dominant themes emerge from their speculation, and what do these tell us about the way illness was perceived in the nineteenth century?

2. Roderick claims his "diseased self-contemplation" feeds the serpent that lives in his bosom. What is "diseased" about it? Based on the evidence from the story, what would you say is the remedy for this disease? Also, health care providers are often urged to be reflective in practice. What is the goal of being reflective, and how does this type of reflection differ from the contemplation in which Roderick engages?

3. Quack doctors of the nineteenth century—like the empiric in this tale— are often depicted as hawking ineffective and even dangerous medicines in lieu of finding real cures. How have our perceptions of drugs evolved over time? What social, economical, and cultural forces have contributed to that evolution? Consider watching clips from a movie like *Love and Other Drugs* (2010) and incorporating that material about the contemporary drug industry into your discussion.

4. Elliston describes the gaze of his sculptor friend, Herkimer, as "the compassionate, yet steady countenance of the sculptor." In what ways does the artist's gaze differ from the "clinical gaze" (Foucault 1994, 108) that stressed objective observation over interpretation? To what extent do the health humanities seek to influence the way practitioners gaze at patients today? For what purpose, for example, might health care professions students analyze works of visual art?

5. What role does Rosina play in Roderick's tale of illness? In what ways does this tale speak to issues related to family dynamics in health and healing?

6. One man asks of Elliston, "Why is this lunatic allowed to go at large?" Does his question reflect nineteenth-century cultural attitudes toward mental illness? Do these attitudes persist today? How do sociocultural attitudes toward mental illness affect the delivery of health care and, ultimately, patient outcomes?

7. Hawthorne's narrator claims, "All persons, chronically diseased, are egotists, whether the disease be of the mind or body." Read the rest of the passage that follows: What does Hawthorne mean when he calls the chronically ill "egotists"? How would you characterize the relationship between illness and identity?

Rappaccini's Daughter

FROM THE WRITINGS OF AUBÉPINE

December 1844, *United States Magazine and Democratic Review*

Under what circumstances does medicine turn to poison? "Rappaccini's Daughter" addresses this question on multiple levels. Literally, Hawthorne was interested in the idea that a human being might become poisonous to others, a concept he likely gleaned from Thomas Browne's Pseudodoxia Epidemica, *book 7, chapter 16 (Turner 1936). Metaphorically, Hawthorne considers the insidious side effects of "treacherously unscientific" pursuits (Uroff 1972, 62), intellectual rivalries, unfounded rumors, and inconstant love. Even an antidote meant to cure can be poisonous under the right (or wrong) circumstances. At the heart of these circumstances stands Beatrice Rappaccini, the eponymous daughter, who embodies the idea that all humans are a mixture of purity and poison. She is a hybrid creature, a fusion of previous female stereotypes: angel and monster, victim and aggressor (Reynolds 1988). Medically speaking, she is poison and remedy, scientific experiment and scientist. But does she have any power?*

Set "very long ago" in Padua, Italy, during the sixteenth-century Italian medical Renaissance period, "Rappaccini's Daughter" reflects the "ideological competition— the vying for legitimation and authority" of rival medical practitioners of the time. Scholars have suggested the rivalry represents that between traditional Galenists and the new "dissecters" like prominent physician Vesalius (Browner 2005; Cerulli and Berry 2014, 113) or between Galenists and those who followed the teachings of the Swiss physician Paracelsus, who rejected traditional teachings and medical authorities in favor of observation and experimentation (Bensick 1985; Stripling 2013).[1] Because of its themes of professional rivalry, the tale offers opportunities to analyze how institutional culture can indirectly affect practice. And since Beatrice Rappaccini allegedly holds a wealth of scientific knowledge, perhaps more than the doctor-scientists at the university, readers might carefully consider her relationship to power and powerlessness: if she has knowledge, what holds her back?

Hawthorne had seen firsthand the potential for medicine to hold women back. The life of his wife, Sophia Hawthorne, another inspiration for the tale, "reads as a virtual encyclopedia of medical poisons and vulnerability of women under pa-

ternal care" (Cerulli and Berry 2014, 119). Her own father, Dr. Nathaniel Peabody, a homeopath and dentist, was the first in a line of irregular and regular doctors to medicate her and was perhaps the original cause her invalidism. According to Julian, the Hawthornes' son, his mother was "incontinently dosed with drugs" (as quoted in Mellow 1980, 133) as a teething infant and never recovered from the effects. Notably, Peabody grew Solanum dulcamara, *a purple wildflower used to relieve toothaches, in his garden (Stripling 2013). Thus, Hawthorne knew personally the perils of nineteenth-century experimentation. Although he sets his tale in sixteenth-century Italy, he arguably indicts nineteenth-century approaches to health that, in the case of women on whom physicians often pioneered experimental remedies and life-threatening surgeries, seemed "decidedly unscientific and even obsessive" (Verbrugge 1976, 960).*

Like "The Birthmark," another tale of enclosure and objectification (Elbert 2004b), "Rappaccini's Daughter" ignites ethical questions related to research on human subjects with or without consent. These questions of power are relevant to any discussion of women's health in the nineteenth century, when the medical establishment ingrained in women the notion of their own physical frailty and inferiority to men. One might consider carefully the expectations placed on nineteenth-century women to be domestic goddesses, angels of the house—and how these expectations were poisonous when they limited women from pursuing mental and physical activities that could have led to better health. As a story that challenges the notion that women should be limited to the domestic sphere (or to any one prescribed sphere), "Rappaccini's Daughter" compares to Charlotte Perkins Gilman's "The Yellow Wallpaper," another tale about medical practices that effectively trap a woman in the domestic world and sap her of the creative powers that could have saved her.[2]

·

WE do not remember to have seen any translated specimens of the productions of M. de l'Aubépine—a fact the less to be wondered at, as his very name is unknown to many of his own countrymen as well as to the student of foreign literature. As a writer, he seems to occupy an unfortunate position between the Transcendentalists (who, under one name or another, have their share in all the current literature of the world) and the great body of pen-and-ink men who address the intellect and sympathies of the multitude. If not too refined, at all events too remote, too shadowy, and unsubstantial in his modes of development to suit the taste of the latter class, and yet too popular to satisfy the spiritual or metaphysical requisitions of the former, he must necessarily find himself without an audience, except here and there an individual or possibly an isolated clique. His writings, to do them justice, are not altogether destitute of fancy and originality; they might have won him greater reputation but for an inveterate love of allegory, which is apt to invest his plots and characters with the aspect of scenery and people in the clouds, and to steal away the human warmth out of his conceptions. His fictions are sometimes historical, sometimes of the present day, and sometimes, so far as can be discovered, have little or no reference either to time or space. In any case, he generally contents himself with a very slight embroidery of outward manners,—the faintest possible counterfeit of real life,—and endeavors to create an interest by some less obvious peculiarity of the subject. Occasionally, a breath of Nature, a raindrop of pathos and tenderness, or a gleam of humor, will find its way into the midst of his fantastic imagery, and make us feel as if, after all, we were yet within the limits of our native earth. We will only add to this very cursory notice that M. de l'Aubépine's productions, if the reader chance to take them in precisely the proper point of view, may amuse a leisure hour as well as those of a brighter man; if otherwise, they can hardly fail to look excessively like nonsense.

Our author is voluminous; he continues to write and publish with as much praiseworthy and indefatigable prolixity as if his efforts were crowned with the brilliant success that so justly attends those of Eugene Sue. His first appearance was by a collection of stories in a long series of volumes, entitled "Comes deux fois racontées." The titles of some of his more recent works (we quote from memory) are as follows: "Le Voyage Céleste à Chemin de Fer," 3 tom. 1838; "Le nouveau Père Adam et la nouvelle Mère Eve," 2 tom. 1839; "Roderic; ou le Serpent à l'estomac," 2 tom. 1840; "Le Culte du Feu," a folio volume of ponderous research into the religion and ritual of the old Persian Ghebers, published in 1841; "La Soirée du Chateau en Espagne," 1 tom. 8vo, 1842; and "L'Artiste du Beau; ou le Papillon Mécanique," 5 tom. 4to, 1843. Our somewhat wearisome perusal of this startling catalogue of volumes has left behind it a certain personal affection and sympathy, though by no means admiration, for M. de l'Aubépine; and we would fain do the little in our power towards introducing him favorably to the American public. The ensuing tale is a translation of his "Beatrice; ou la Belle Empoisonneuse," recently published in "La Revue Anti-Aristocratique." This journal, edited by the Comte de Bearhaven, has, for some years past, led the defence of liberal principles and popular rights, with a faithfulness and ability worthy of all praise.

A young man, named Giovanni Guasconti, came, very long ago, from the more southern region of Italy, to pursue his studies at the University of Padua. Giovanni, who had but a scanty supply of gold ducats in his pocket, took lodgings in a high and gloomy chamber of an old edifice which looked not unworthy to have been the palace of a Paduan noble, and which, in fact, exhibited over its entrance the armorial bearings of a family long since extinct. The young stranger, who was not unstudied in the great poem of his country, recollected that one of the ancestors of this family, and perhaps an occupant of this very mansion, had been pictured by Dante as a partaker of the immortal agonies of his Inferno. These reminiscences and associations, together with the tendency to heartbreak natural to a young man for the first time out of his native sphere, caused Giovanni to sigh heavily as he looked around the desolate and ill-furnished apartment.

"Holy Virgin, signor," cried old Dame Lisabetta, who, won by the youth's remarkable beauty of person, was kindly endeavoring to give the chamber a habitable air, "what a sigh was that to come out of a young man's heart! Do you find this old mansion gloomy? For the love of Heaven, then, put your head out of the window, and you will see as bright sunshine as you have left in Naples."

Guasconti mechanically did as the old woman advised, but could not quite agree with her that the Paduan sunshine was as cheerful as that of southern Italy. Such as it was, however, it fell upon a garden beneath the window and expended its fostering influences on a variety of plants, which seemed to have been cultivated with exceeding care.

"Does this garden belong to the house?" asked Giovanni.

"Heaven forbid, signor, unless it were fruitful of better pot herbs than any that grow there now," answered old Lisabetta. "No; that garden is cultivated by the own hands of Signor Giacomo Rappaccini, the famous doctor, who, I warrant him, has been heard of as far as Naples. It is said he distils these plants into medicines that are as potent as a charm. Oftentimes you may see the signor doctor at work, and perchance the signora, his daughter, too, gathering the strange flowers that grow in the garden."

The old woman had now done what she could for the aspect of the chamber; and, commending the young man to the protection of the saints, took her departure.

Giovanni still found no better occupation than to look down into the garden beneath his window. From its appearance, he judged it to be one of those botanic gardens which were of earlier date in Padua than elsewhere in Italy or in the world. Or, not improbably, it might once have been the pleasure-place of an opulent family; for there was the ruin of a marble fountain in the centre, sculptured with rare art, but so wofully shattered that it was impossible to trace the original design from the chaos of remaining fragments. The water, however, continued to gush and sparkle into the sunbeams as cheerfully as ever. A little gurgling sound ascended to the young man's window, and made him feel as if a fountain were an immortal spirit that sung its song unceasingly and without heeding the vicissitudes around it, while one century imbodied it in marble and another scattered the perishable garniture on the soil. All about the pool into which the water subsided grew various plants, that seemed to require a plentiful supply of moisture for the nourishment of gigantic leaves, and, in some instances, flowers gorgeously magnificent. There was one shrub in particular, set in a marble vase in the midst of the pool, that bore a profusion of purple blossoms, each of which had the lustre and richness of a gem; and the whole together made a show so resplendent that it seemed enough to illuminate the garden, even had there been no sunshine. Every portion of the soil was peopled with plants and herbs, which, if less beautiful, still bore tokens of assiduous care, as if all had their individual virtues, known to the scientific mind that fostered them. Some were placed in urns, rich with old carving, and others in common garden pots; some crept serpent-like along the ground, or climbed on high, using whatever means of ascent was offered them. One plant had wreathed itself round a statue of Vertumnus, which was thus quite veiled and shrouded in a drapery of hanging foliage, so happily arranged that it might have served a sculptor for a study.

While Giovanni stood at the window he heard a rustling behind a screen of leaves, and became aware that a person was at work in the garden. His figure soon emerged into view, and showed itself to be that of no common laborer, but a tall, emaciated, sallow, and sickly-looking man, dressed in a scholar's garb of black. He was beyond the middle term of life, with gray hair, a thin, gray beard, and a face singularly marked with intellect and cultivation, but which could never, even in his more youthful days, have expressed much warmth of heart.

Nothing could exceed the intentness with which this scientific gardener examined every shrub which grew in his path: it seemed as if he was looking into their inmost nature, making observations in regard to their creative essence, and discovering why one leaf grew in this shape and another in that, and wherefore such and such flowers differed among themselves in hue and perfume. Nevertheless, in spite of the deep intelligence on his part, there was no approach to intimacy between himself and these vegetable existences. On the contrary, he avoided their actual touch or the direct inhaling of their odors with a caution that impressed Giovanni most disagreeably; for the man's demeanor was that of one walking among malignant influences, such as savage beasts, or deadly snakes, or evil spirits, which, should he allow them one moment of license, would wreak upon him some terrible fatality. It was strangely frightful to the young man's imagination, to see this air of insecurity in a person cultivating a garden, that most simple and innocent of human toils, and which had been alike the joy and labor of the unfallen parents of the race. Was this garden, then, the Eden of the present world? And this man, with such a perception of harm in what his own hands caused to grow,—was he the Adam?

The distrustful gardener, while plucking away the dead leaves or pruning the too luxuriant growth of the shrubs, defended his hands with a pair of thick gloves. Nor were these his only armor. When, in his walk through the garden, he came to the magnificent plant that hung its purple gems beside the marble fountain, he placed a kind of mask over his mouth and nostrils, as if all this beauty did but conceal a deadlier malice; but, finding his task still too dangerous, he drew back, removed the mask, and called loudly, but in the infirm voice of a person affected with inward disease,—

"Beatrice! Beatrice!"

"Here am I, my father! What would you?" cried a rich and youthful voice from the window of the opposite house—a voice as rich as a tropical sunset, and which made Giovanni, though he knew not why, think of deep hues of purple or crimson and of perfumes heavily delectable. "Are you in the garden?"

"Yes, Beatrice," answered the gardener, "and I need your help."

Soon there emerged from under a sculptured portal the figure of a young girl, arrayed with as much richness of taste as the most splendid of the flowers, beautiful as the day, and with a bloom so deep and vivid that one shade more would have been too much. She looked redundant with life, health, and energy; all of which attributes were bound down and compressed, as it were, and girdled tensely, in their luxuriance, by her virgin zone. Yet Giovanni's fancy must have grown morbid, while he looked down into the garden; for the impression which the fair stranger made upon him was as if here were another flower, the human sister of those vegetable ones, as beautiful as they, more beautiful than the richest of them, but still to be touched only with a glove, nor to be approached without a mask. As Beatrice came down the garden path, it was observable that she handled and inhaled the odor of several of the plants, which her father had most sedulously avoided.

"Here, Beatrice," said the latter, "see how many needful offices require to be done to our chief treasure. Yet, shattered as I am, my life might pay the penalty of approaching it so closely as circumstances demand. Henceforth, I fear, this plant must be consigned to your sole charge."

"And gladly will I undertake it," cried again the rich tones of the young lady, as she bent towards the magnificent plant and opened her arms as if to embrace it. "Yes, my sister, my splendor, it shall be Beatrice's task to nurse and serve thee; and thou shalt reward her with thy kisses and perfumed breath, which to her is as the breath of life."

Then, with all the tenderness in her manner that was so strikingly expressed in her words, she busied herself with such attentions as the plant seemed to require; and Giovanni, at his lofty window, rubbed his eyes and almost doubted whether it were a girl tending her favorite flower, or one sister performing the duties of affection to another. The scene soon terminated. Whether Dr. Rappaccini had finished his labors in the garden, or that his watchful eye had caught the stranger's face, he now took his daughter's arm and retired. Night was already closing in; oppressive exhalations seemed to proceed from the plants and steal upward past the open window; and Giovanni, closing the lattice, went to his couch and dreamed of a rich flower and beautiful girl. Flower and maiden were different, and yet the same, and fraught with some strange peril in either shape.

But there is an influence in the light of morning that tends to rectify whatever errors of fancy, or even of judgment, we may have incurred during the sun's decline, or among the shadows of the night, or in the less wholesome glow of moonshine. Giovanni's first movement on starting from sleep, was to throw open the window and gaze down into the garden which his dreams had made so fertile of mysteries. He was surprised and a little ashamed to find how real and matter-of-fact an affair it proved to be, in the first rays of the sun which gilded the dew-drops that hung upon leaf and blossom, and, while giving a brighter beauty to each rare flower, brought everything within the limits of ordinary experience. The young man rejoiced, that, in the heart of the barren city, he had the privilege of overlooking this spot of lovely and luxuriant vegetation. It would serve, he said to himself, as a symbolic language, to keep him in communion with Nature. Neither the sickly and thoughtworn Dr. Giacomo Rappaccini, it is true, nor his brilliant daughter, were now visible; so that Giovanni could not determine how much of the singularity which he attributed to both was due to their own qualities and how much to his wonder-working fancy; but he was inclined to take a most rational view of the whole matter.

In the course of the day he paid his respects to Signor Pietro Baglioni, professor of medicine in the university, a physician of eminent repute, to whom Giovanni had brought a letter of introduction. The professor was an elderly personage, apparently of genial nature, and habits that might almost be called jovial. He kept the young man to dinner, and made himself very agreeable by the freedom and liveliness of his conversation, especially when warmed by a flask or two of Tuscan wine. Giovanni, conceiving that men of science, inhabitants of the same city, must needs be on familiar terms with one another, took an opportunity to mention the name of Dr. Rappaccini. But the professor did not respond with so much cordiality as he had anticipated.

"Ill would it become a teacher of the divine art of medicine," said Professor Pietro Baglioni, in answer to a question of Giovanni, "to withhold due and well-considered praise of a physician so eminently skilled as Rappaccini; but, on the other hand, I should answer it but scantily to my conscience were I to permit a worthy youth like yourself, Signor Giovanni, the son of an ancient friend, to imbibe erroneous ideas respecting a man who might hereafter chance to hold your life and death in his hands. The truth is, our worshipful Dr. Rappaccini has as much science as any member of the faculty—with perhaps one single exception—in Padua, or all Italy; but there are certain grave objections to his professional character."

"And what are they?" asked the young man.

"Has my friend Giovanni any disease of body or heart, that he is so inquisitive about physicians?" said the professor, with a smile. "But as for Rappaccini, it is said of him—and I, who know the man well, can answer for its truth—that he cares infinitely more for science than for mankind. His patients are interesting to him only as subjects for some new experiment. He would sacrifice human life, his own among the rest, or whatever else was dearest to him, for the sake of adding so much as a grain of mustard seed to the great heap of his accumulated knowledge."

"Methinks he is an awful man indeed," remarked Guasconti, mentally recalling the cold and purely intellectual aspect of Rappaccini. "And yet, worshipful professor, is it not a noble spirit? Are there many men capable of so spiritual a love of science?"

"God forbid," answered the Professor, somewhat testily; "at least, unless they take sounder views of the healing art than those adopted by Rappaccini. It is his theory that all medicinal virtues are comprised within those substances which we term vegetable poisons. These he cultivates with his own hands, and is said even to have produced new varieties of poison, more horribly deleterious than Nature, without the assistance of this learned person, would ever have plagued the world withal. That the signor doctor does less mischief than might be expected with such dangerous substances is undeniable. Now and then, it must be owned, he has effected, or seemed to effect, a marvellous cure; but, to tell you my private mind, Signor Giovanni, he should receive little credit for such instances of success,—they being probably the work of chance,—but should be held strictly accountable for his failures, which may justly be considered his own work."

The youth might have taken Baglioni's opinions with many grains of allowance had he known that there was a professional warfare of long continuance between him and Dr. Rappaccini, in which the latter was generally thought to have gained the advantage. If the reader be inclined to judge for himself, we refer him to certain black-letter tracts on both sides, preserved in the medical department of the University of Padua.

"I know not, most learned professor," returned Giovanni, after musing on what had been said of Rappaccini's exclusive zeal for science,—"I know not how dearly this physician may love his art; but surely there is one object more dear to him. He has a daughter."

"Aha!" cried the professor, with a laugh. "So now our friend Giovanni's secret is out. You have heard of this daughter, whom all the young men in Padua are wild about, though not half a dozen have ever had the good hap to see her face. I know little of the Signora Beatrice, save that Rappaccini is said to have instructed her deeply in his science, and that, young and beautiful as fame reports her, she is already qualified to fill a professor's chair. Perchance her father destines her for mine! Other absurd rumors there be, not worth talking about or listening to. So now, Signor Giovanni, drink off your glass of lachryma."

Guasconti returned to his lodgings somewhat heated with the wine he had quaffed, and which caused his brain to swim with strange fantasies in reference to Dr. Rappaccini and the beautiful Beatrice. On his way, happening to pass by a florist's, he bought a fresh bouquet of flowers.

Ascending to his chamber, he seated himself near the window, but within the shadow thrown by the depth of the wall, so that he could look down into the garden with little risk of being discovered. All beneath his eye was a solitude. The strange plants were basking in the sunshine, and now and then nodding gently to one another, as if in acknowledgment of sympathy and kindred. In the midst, by the shattered fountain, grew the magnificent shrub, with its purple gems clustering all over it; they glowed in the air, and gleamed back again out of the depths of the pool, which thus seemed to overflow with colored radiance from the rich reflection that was steeped in it. At first, as we have said, the garden was a solitude. Soon, however,—as Giovanni had half hoped, half feared, would be the case,—a figure appeared beneath the antique sculptured portal, and came down between the rows of plants, inhaling their various perfumes, as if she were one of those beings of old classic fable, that lived upon sweet odors. On again beholding Beatrice, the young man was even startled to perceive how much her beauty exceeded his recollection of it; so brilliant, so vivid, in its character, that she glowed amid the sunlight, and, as Giovanni whispered to himself, positively illuminated the more shadowy intervals of the garden path. Her face being now more revealed than on the former occasion, he was struck by its expression of simplicity and sweetness,—qualities that had not entered into his idea of her character, and which made him ask anew, what manner of mortal she might be. Nor did he fail again to observe, or imagine, an analogy between the beautiful girl and the gorgeous shrub that hung its gemlike flowers over the fountain,—a resemblance which Beatrice seemed to have indulged a fantastic humor in heightening, both by the arrangement of her dress and the selection of its hues.

Approaching the shrub, she threw open her arms, as with a passionate ardor, and drew its branches into an intimate embrace—so intimate that her features were hidden in its leafy bosom and her glistening ringlets all intermingled with the flowers.

"Give me thy breath, my sister," exclaimed Beatrice; "for I am faint with common air. And give me this flower of thine, which I separate with gentlest fingers from the stem and place it close beside my heart."

With these words, the beautiful daughter of Rappaccini plucked one of the richest blossoms of the shrub, and was about to fasten it in her bosom. But now, unless Giovanni's draughts of wine had bewildered his senses, a singular incident occurred. A small orange colored reptile, of the lizard or chameleon species, chanced to be creeping along the path, just at the feet of Beatrice. It appeared to Giovanni,—but, at the distance from which he gazed, he could scarcely have seen anything so minute,—it appeared to him, however, that a drop or two of moisture from the broken stem of the flower descended upon the lizard's head. For an instant, the reptile contorted itself violently, and then lay motionless in the sunshine. Beatrice observed this remarkable phenomenon, and crossed herself, sadly, but without surprise; nor did she therefore hesitate to arrange the fatal flower in her bosom. There it blushed, and almost glimmered with the dazzling effect of a precious stone, adding to her dress and aspect the one appropriate charm which nothing else in the world could have supplied. But Giovanni, out of the shadow of his window, bent forward and shrank back, and murmured and trembled.

"Am I awake? Have I my senses?" said he to himself. "What is this being? Beautiful shall I call her, or inexpressibly terrible?"

Beatrice now strayed carelessly through the garden, approaching closer beneath Giovanni's window, so that he was compelled to thrust his head quite out of its concealment in order to gratify the intense and painful curiosity which she excited. At this moment there came a beautiful insect over the garden wall; it had, perhaps, wandered through the city, and found no flowers or verdure among those antique haunts of men until the heavy perfumes of Dr. Rappaccini's shrubs had lured it from afar. Without alighting on the flowers, this winged brightness seemed to be attracted by Beatrice, and lingered in the air and fluttered about her head. Now, here it could not be but that Giovanni

Guasconti's eyes deceived him. Be that as it might, he fancied that, while Beatrice was gazing at the insect with childish delight, it grew faint and fell at her feet; its bright wings shivered; it was dead—from no cause that he could discern, unless it were the atmosphere of her breath. Again Beatrice crossed herself and sighed heavily as she bent over the dead insect.

An impulsive movement of Giovanni drew her eyes to the window. There she beheld the beautiful head of the young man—rather a Grecian than an Italian head, with fair, regular features, and a glistening of gold among his ringlets— gazing down upon her like a being that hovered in mid air. Scarcely knowing what he did, Giovanni threw down the bouquet which he had hitherto held in his hand.

"Signora," said he, "there are pure and healthful flowers. Wear them for the sake of Giovanni Guasconti."

"Thanks, signor," replied Beatrice, with her rich voice, that came forth as it were like a gush of music, and with a mirthful expression half childish and half woman-like. "I accept your gift, and would fain recompense it with this precious purple flower; but if I toss it into the air it will not reach you. So Signor Guasconti must even content himself with my thanks."

She lifted the bouquet from the ground, and then, as if inwardly ashamed at having stepped aside from her maidenly reserve to respond to a stranger's greeting, passed swiftly homeward through the garden. But few as the moments were, it seemed to Giovanni, when she was on the point of vanishing beneath the sculptured portal, that his beautiful bouquet was already beginning to wither in her grasp. It was an idle thought; there could be no possibility of distinguishing a faded flower from a fresh one at so great a distance.

For many days after this incident the young man avoided the window that looked into Dr. Rappaccini's garden, as if something ugly and monstrous would have blasted his eyesight had he been betrayed into a glance. He felt conscious of having put himself, to a certain extent, within the influence of an unintelligible power by the communication which he had opened with Beatrice. The wisest course would have been, if his heart were in any real danger, to quit his lodgings and Padua itself, at once; the next wiser, to have accustomed himself, as far as possible, to the familiar and daylight view of Beatrice—thus bringing her rigidly and systematically within the limits of ordinary experience. Least of all, while avoiding her sight, ought Giovanni have remained so near this extraordinary being, that the proximity and possibility even of intercourse should give a kind of substance and reality to the wild vagaries which his imagination ran riot continually in producing. Guasconti had not a deep heart—or at all events, its depths were not sounded now; but he had a quick fancy, and an ardent southern temperament, which rose every instant to a higher fever pitch. Whether or no Beatrice possessed those terrible attributes, that fatal breath, the affinity with those so beautiful and deadly flowers which were indicated by what Giovanni had witnessed, she had at least instilled a fierce and subtle poison into his system. It was not love, although her rich beauty was a madness to him; nor horror, even while he fancied her spirit to be imbued with the same baneful essence that seemed to pervade her physical frame; but a wild offspring of both love and horror that had each parent in it, and burned like one and shivered like the other. Giovanni knew not what to dread; still less did he know what to hope; yet hope and dread kept a continual warfare in his breast, alternately vanquishing one another and starting up afresh to renew the contest. Blessed are all simple emotions, be they dark or bright! It is the lurid intermixture of the two that produces the illuminating blaze of the infernal regions.

Sometimes he endeavored to assuage the fever of his spirit by a rapid walk through the streets of Padua or beyond its gates: his footsteps kept time with the throbbings of his brain, so that the walk was apt to accelerate itself to a race. One day he found himself arrested; his arm was seized by a portly personage, who had turned back on recognizing the young man and expended much breath in overtaking him.

"Signor Giovanni! Stay, my young friend!" cried he. "Have you forgotten me? That might well be the case, if I were as much altered as yourself."

It was Baglioni, whom Giovanni had avoided, ever since their first meeting, from a doubt that the professor's sagacity would look too deeply into his secrets. Endeavoring to recover himself, he stared forth wildly from his inner world into the outer one and spoke like a man in a dream.

"Yes; I am Giovanni Guasconti. You are Professor Pietro Baglioni. Now let me pass!"

"Not yet, not yet, Signor Giovanni Guasconti," said the professor, smiling, but at the same time scrutinizing the youth with an earnest glance. "What! did I grow up side by side with your father? and shall his son pass me like a stranger, in these old streets of Padua? Stand still, Signor Giovanni; for we must have a word or two before we part."

"Speedily, then, most worshipful professor, speedily," said Giovanni, with feverish impatience. "Does not your worship see that I am in haste?"

Now, while he was speaking there came a man in black along the street, stooping and moving feebly, like a person in inferior health. His face was all overspread with a most sickly and sallow hue, but yet so pervaded with an expression of piercing and active intellect that an observer might easily have overlooked the merely physical attributes, and have seen only this wonderful energy. As he passed, this person exchanged a cold and distant salutation with Baglioni, but fixed his eyes upon Giovanni with an intentness that seemed to bring out whatever was within him worthy of notice. Nevertheless, there was a peculiar quietness in the look, as if taking merely a speculative, not a human interest, in the young man.

"It is Dr. Rappaccini!" whispered the professor when the stranger had passed. "Has he ever seen your face before?"

"Not that I know," answered Giovanni, starting at the name.

"He *has* seen you! he must have seen you!" said Baglioni, hastily. "For some purpose or other, this man of science is making a study of you. I know that look of his! It is the same that coldly illuminates his face as he bends over a bird, a mouse, or a butterfly, which, in pursuance of some experiment, he has killed by the perfume of a flower; a look as deep as Nature itself, but without Nature's warmth of love. Signor Giovanni, I will stake my life upon it, you are the subject of one of Rappaccini's experiments!"

"Will you make a fool of me?" cried Giovanni, passionately. "*That,* signor professor, were an untoward experiment."

"Patience! patience!" replied the imperturbable professor. "I tell thee, my poor Giovanni, that Rappaccini has a scientific interest in thee. Thou hast fallen into fearful hands! And the Signora Beatrice,—what part does she act in this mystery?"

But Guasconti, finding Baglioni's pertinacity intolerable, here broke away, and was gone before the professor could again seize his arm. He looked after the young man intently and shook his head.

"This must not be," said Baglioni to himself. "The youth is the son of my old friend, and shall not come to any harm from which the arcana of medical science can preserve him. Besides, it is too insufferable an impertinence in Rappaccini thus to snatch the lad out of my own hands, as I may say, and make use of him for his infernal experiments. This daughter of his! It shall be looked to. Perchance, most learned Rappaccini, I may foil you where you little dream of it!"

Meanwhile Giovanni had pursued a circuitous route, and at length found himself at the door of his lodgings. As he crossed the threshold he was met by old Lisabetta, who smirked and smiled, and was evidently desirous to attract his attention; vainly, however, as the ebullition of his feelings had momentarily subsided into a cold and dull vacuity. He turned his eyes full upon the withered face that was puckering itself into a smile, but seemed to behold it not. The old dame, therefore, laid her grasp upon his cloak.

"Signor! signor!" whispered she, still with a smile over the whole breadth of her visage, so that it looked not unlike a grotesque carving in wood, darkened by centuries. "Listen, signor! There is a private entrance into the garden!"

"What do you say?" exclaimed Giovanni, turning quickly about, as if an inanimate thing should start into feverish life. "A private entrance into Dr. Rappaccini's garden?"

"Hush! hush! not so loud!" whispered Lisabetta, putting her hand over his mouth. "Yes; into the worshipful doctor's garden, where you may see all his fine shrubbery. Many a young man in Padua would give gold to be admitted among those flowers."

Giovanni put a piece of gold into her hand.

"Show me the way," said he.

A surmise, probably excited by his conversation with Baglioni, crossed his mind, that this interposition of old Lisabetta might perchance be connected with the intrigue, whatever were its nature, in which the professor seemed to suppose that Dr. Rappaccini was involving him. But such a suspicion, though it disturbed Giovanni, was inadequate to restrain him. The instant he was aware of the possibility of approaching Beatrice, it seemed an absolute necessity of his existence to do so. It mattered not whether she were angel or demon; he was irrevocably within her sphere, and must obey the law that whirled him onward, in ever-lessening circles, towards a result which he did not attempt to foreshadow; and yet, strange to say, there came across him a sudden doubt whether this intense interest on his part were not delusory; whether it were really of so deep and positive a nature as to justify him in now thrusting himself into an incalculable position; whether it were not merely the fantasy of a young man's brain, only slightly or not at all, connected with his heart.

He paused, hesitated, turned half about, but again went on. His withered guide led him along several obscure passages, and finally undid a door, through which, as it was opened, there came the sight and sound of rustling leaves, with the broken sunshine glimmering among them. Giovanni stepped forth, and forcing himself through the entanglement of a shrub that wreathed its tendrils over the hidden entrance, stood beneath his own window, in the open area of Dr. Rappaccini's garden.

How often is it the case that, when impossibilities have come to pass and dreams have condensed their misty substance into tangible realities, we find ourselves calm, and even coldly self-possessed, amid circumstances which it would have been a delirium of joy or agony to anticipate! Fate delights to thwart us thus. Passion will choose his own time to rush upon the scene, and lingers sluggishly behind when an appropriate adjustment of events would seem to summon his appearance. So was it now with Giovanni. Day after day his pulses had throbbed with feverish blood, at the improbable idea of an interview with Beatrice, and of standing with her, face to face, in this very garden, basking in the Oriental sunshine of her beauty, and snatching from her full gaze the mystery which he deemed the riddle of his own existence. But now there was a singular and untimely equanimity within his breast. He threw a glance around the garden to discover if Beatrice or her father were present, and, perceiving that he was alone, began a critical observation of the plants.

The aspect of one and all of them dissatisfied him; their gorgeousness seemed fierce, passionate, and even unnatural. There was hardly an individual shrub which a wanderer, straying by himself through a forest, would not have been startled to find growing wild, as if an unearthly face had glared at him out of the thicket. Several also would have shocked a delicate instinct by an appearance of artificiality indicating that there had been such commixture, and, as it were, adultery of various vegetable species, that the production was no longer of God's making, but the monstrous offspring of man's depraved fancy, glowing with only an evil mockery of beauty. They were probably the result of experiment, which in one or two cases had succeeded in mingling plants individually lovely into a compound possessing the questionable and ominous character that distinguished the whole growth of the garden. In fine, Giovanni recognized but two or three plants in the collection, and those of a kind that he well knew to be poisonous. While busy with these contemplations he heard the rustling of a silken garment, and, turning, beheld Beatrice emerging from beneath the sculptured portal.

Giovanni had not considered with himself what should be his deportment; whether he should apologize for his intrusion into the garden, or assume that he was there with the privity at least, if not by the desire, of Dr. Rappaccini or his daughter; but Beatrice's manner placed him at his ease, though leaving him still in doubt by what agency he had gained admittance. She came lightly along the path and met him near the broken fountain. There was surprise in her face, but brightened by a simple and kind expression of pleasure.

"You are a connoisseur in flowers, Signor," said Beatrice with a smile, alluding to the bouquet which he had flung her from the window. "It is no marvel, therefore, if the sight of my father's rare collection has tempted you to take a nearer view. If he were here, he could tell you many strange and interesting facts as to the nature and habits of these shrubs; for he has spent a lifetime in such studies, and this garden is his world."

"And yourself, lady," observed Giovanni, "if fame says true,—you likewise are deeply skilled in the virtues indicated by these rich blossoms and these spicy perfumes. Would you deign to be my instructress, I should prove an apter scholar than if taught by Signor Rappaccini himself."

"Are there such idle rumors?" asked Beatrice, with the music of a pleasant laugh. "Do people say that I am skilled in my father's science of plants? What a jest is there! No; though I have grown up among these flowers, I know no more of them than their hues and perfume; and sometimes, methinks I would fain rid myself of even that small knowledge. There are many flowers here, and those not the least brilliant, that shock and offend me, when they meet my eye. But, pray, signor, do not believe these stories about my science. Believe nothing of me save what you see with your own eyes."

"And must I believe all that I have seen with my own eyes?" asked Giovanni, pointedly, while the recollection of former scenes made him shrink. "No, signora; you demand too little of me. Bid me believe nothing save what comes from your own lips."

It would appear that Beatrice understood him. There came a deep flush to her cheek; but she looked full into Giovanni's eyes, and responded to his gaze of uneasy suspicion with a queenlike haughtiness.

"I do so bid you, signor!" she replied. "Forget whatever you may have fancied in regard to me. If true to the outward senses, still it may be false in its essence; but the words of Beatrice Rappaccini's lips are true from the heart outward. Those you may believe."

A fervor glowed in her whole aspect and beamed upon Giovanni's consciousness like the light of truth itself; but while she spoke there was a fragrance in the atmosphere around her, rich and delightful, though evanescent, yet which the young man, from an indefinable reluctance, scarcely dared to draw into his lungs. It might be the odor of the flowers. Could it be Beatrice's breath which thus embalmed her words with a strange richness, as if by steeping them in her heart? A faintness passed like a shadow over Giovanni, and flitted away; he seemed to gaze through the beautiful girl's eyes into her transparent soul, and felt no more doubt or fear.

The tinge of passion that had colored Beatrice's manner vanished; she became gay, and appeared to derive a pure delight from her communion with the youth, not unlike what the maiden of a lonely island might have felt conversing with a voyager from the civilized world. Evidently her experience of life had been confined within the limits of that garden. She talked now about matters as simple as the daylight or summer clouds, and now asked questions in reference to the city, or Giovanni's distant home, his friends, his mother, and his sisters—questions indicating such seclusion, and such lack of familiarity with modes and forms, that Giovanni responded as if to an infant. Her spirit gushed out before him like a fresh rill that was just catching its first glimpse of the sunlight and wondering at the reflections of earth and sky which were flung into its bosom. There came thoughts, too, from a deep source, and fantasies of a gemlike brilliancy, as if diamonds and rubies sparkled upward among the bubbles of the fountain. Ever and anon there gleamed across the young man's mind a sense of wonder that he should be walking side by side with the being who had so wrought upon his imagination, whom he had idealized in such hues of terror, in whom he had positively witnessed such manifestations of dreadful attributes,—that he should be conversing with Beatrice like a brother, and should find her so human and so maidenlike. But such reflections were only momentary; the effect of her character was too real, not to make itself familiar at once.

In this free intercourse they had strayed through the garden, and now, after many turns among its avenues, were come to the shattered fountain, beside which grew the magnificent shrub with its treasury of glowing blossoms. A fragrance was diffused from it which Giovanni recognized as identical with that which he had attributed to Beatrice's breath, but incomparably more powerful. As her eyes fell upon it, Giovanni beheld her press her hand to her bosom as if her heart were throbbing suddenly and painfully.

"For the first time in my life," murmured she, addressing the shrub, "I had forgotten thee."

"I remember, signora," said Giovanni, "that you once promised to reward me with one of these living gems for the bouquet which I had the happy boldness to fling to your feet. Permit me now to pluck it as a memorial of this interview."

He made a step towards the shrub with extended hand; but Beatrice darted forward, uttering a shriek that went through his heart like a dagger. She caught his hand and drew it back with the whole force of her slender figure. Giovanni felt her touch thrilling through his fibres.

"Touch it not!" exclaimed she, in a voice of agony. "Not for thy life! It is fatal!"

Then, hiding her face, she fled from him and vanished beneath the sculptured portal. As Giovanni followed her with his eyes, he beheld the emaciated figure and pale intelligence of Dr. Rappaccini, who had been watching the scene, he knew not how long, within the shadow of the entrance.

No sooner was Guasconti alone in his chamber than the image of Beatrice came back to his passionate musings, invested with all the witchery that had been gathering around it ever since his first glimpse of her, and now likewise imbued with a tender warmth of girlish womanhood. She was human; her nature was endowed with all gentle and feminine qualities; she was worthiest to be worshipped; she was capable, surely, on her part, of the height and heroism of love. Those tokens which he had hitherto considered as proofs of a frightful peculiarity in her physical and moral system were now either forgotten, or, by the subtle sophistry of passion transmitted into a golden crown of enchantment, rendering Beatrice the more admirable by so much as she was the more unique. Whatever had looked ugly was now beautiful; or, if incapable of such a change, it stole away and hid itself among those shapeless half ideas which throng the dim region beyond the daylight of our perfect consciousness. Thus did Giovanni spend the night, nor fell asleep until the dawn had begun to awake the slumbering flowers in Dr. Rappaccini's garden, whither Giovanni's dreams doubtless led him. Up rose the sun in his due season, and, flinging his beams upon the young man's eyelids, awoke him to a sense of pain. When thoroughly aroused, he became sensible of a burning and tingling agony in his hand—in his right hand—the very hand which Beatrice had grasped in her own when he was on the point of plucking one of the gemlike flowers. On the back of that hand there was now a purple print like that of four small fingers, and the likeness of a slender thumb upon his wrist.

Oh, how stubbornly does love,—or even that cunning semblance of love which flourishes in the imagination, but strikes no depth of root into the heart,—how stubbornly does it hold its faith, until the moment comes when it is doomed to vanish into thin mist! Giovanni wrapped a handkerchief about his hand and wondered what evil thing had stung him, and soon forgot his pain in a reverie of Beatrice.

After the first interview, a second was in the inevitable course of what we call fate. A third; a fourth; and a meeting with Beatrice in the garden was no longer an incident in Giovanni's daily life, but the whole space in which he might be said to live; for the anticipation and memory of that ecstatic hour made up the remainder. Nor was it otherwise with the daughter of Rappaccini. She watched for the youth's appearance, and flew to his side with confidence as unreserved as if they had been playmates from early infancy—as if they were such playmates still. If, by any unwonted chance, he failed to come at the appointed moment, she stood beneath the window and sent up the rich sweetness of her tones to float around him in his chamber, and echo and reverberate throughout his heart: "Giovanni! Giovanni! Why tarriest thou? Come down!" And down he hastened into that Eden of poisonous flowers.

But, with all this intimate familiarity, there was still a reserve in Beatrice's demeanor, so rigidly and invariably sustained that the idea of infringing it scarcely occurred to his imagination. By all appreciable signs, they loved; they had looked love with eyes that conveyed the holy secret from the depths of one soul into the depths of the other, as if it were too sacred to be whispered by the way; they had even spoken love in those gushes of passion when their spirits darted forth in articulated breath like tongues of long-hidden flame; and yet there had been no seal of lips, no clasp of hands, nor any slightest caress such as love claims and hallows. He had never touched one of the gleaming ringlets of her hair; her garment—so marked was the physical barrier between them—had never been waved against him by a breeze. On the few occasions when Giovanni had seemed tempted to overstep the limit, Beatrice grew so sad, so stern, and withal wore such a look of desolate separation, shuddering at itself, that not a spoken word was requisite to repel him. At such times he was startled at the horrible suspicions that rose, monster-like, out of the caverns of his heart and stared him in the face; his love grew thin and faint as the morning mist, his doubts alone had substance. But, when Beatrice's face brightened again after the momentary shadow, she was transformed at once from the mysterious, questionable being, whom he had watched with so much awe and horror; she was now the beautiful and unsophisticated girl whom he felt that his spirit knew with a certainty beyond all other knowledge.

A considerable time had now passed since Giovanni's last meeting with Baglioni. One morning, however, he was disagreeably surprised by a visit from the professor, whom he had scarcely thought of for whole weeks, and would willingly have forgotten still longer. Given up as he had long been to a pervading excitement, he could tolerate no companions except upon condition of their perfect sympathy with his present state of feeling. Such sympathy was not to be expected from Professor Baglioni.

The visitor chatted carelessly for a few moments about the gossip of the city and the university, and then took up another topic.

"I have been reading an old classic author lately," said he, "and met with a story that strangely interested me. Possibly you may remember it. It is of an Indian prince, who sent a beautiful woman as a present to Alexander the Great. She was as lovely as the dawn, and gorgeous as the sunset; but what especially distinguished her was a certain rich perfume in her breath—richer than a garden of Persian roses. Alexander, as was natural to a youthful conqueror, fell in love at first sight with this magnificent stranger; but a certain sage physician, happening to be present, discovered a terrible secret in regard to her."

"And what was that?" asked Giovanni, turning his eyes downward to avoid those of the professor.

"That this lovely woman," continued Baglioni, with emphasis, "had been nourished with poisons from her birth upward, until her whole nature was so imbued with them that she herself had become the deadliest poison in existence. Poison was her element of life. With that rich perfume of her breath, she blasted the very air. Her love would have been poison—her embrace death. Is not this a marvellous tale?"

"A childish fable," answered Giovanni, nervously starting from his chair. "I marvel how your worship finds time to read such nonsense among your graver studies."

"By the by," said the professor, looking uneasily about him, "what singular fragrance is this in your apartment? Is it the perfume of your gloves? It is faint, but delicious; and yet, after all, by no means agreeable. Were I to breathe it long, methinks it would make me ill. It is like the breath of a flower; but I see no flowers in the chamber."

"Nor are there any," replied Giovanni, who had turned pale as the professor spoke; "nor, I think, is there any fragrance, except in your worship's imagination. Odors, being a sort of element combined of the sensual and the spiritual, are apt to deceive us in this manner. The recollection of a perfume, the bare idea of it, may easily be mistaken for a present reality."

"Ay; but my sober imagination does not often play such tricks," said Baglioni; "and, were I to fancy any kind of odor, it would be that of some vile apothecary drug, wherewith my fingers are likely enough to be imbued. Our worshipful friend Rappaccini, as I have heard, tinctures his medicaments with odors richer than those of Araby. Doubtless, likewise, the fair and learned Signora Beatrice would minister to her patients with draughts as sweet as a maiden's breath. But woe to him that sips them!"

Giovanni's face evinced many contending emotions. The tone in which the professor alluded to the pure and lovely daughter of Rappaccini was a torture to his soul; and yet the intimation of a view of her character, opposite to his own, gave instantaneous distinctness to a thousand dim suspicions, which now grinned at him like so many demons. But he strove hard to quell them and to respond to Baglioni with a true lover's perfect faith.

"Signor professor," said he, "you were my father's friend; perchance, too, it is your purpose to act a friendly part towards his son. I would fain feel nothing towards you save respect and deference; but I pray you to observe, signor, that there is one subject on which we must not speak. You know not the Signora Beatrice. You cannot, therefore, estimate the wrong—the blasphemy, I may even say—that is offered to her character by a light or injurious word."

"Giovanni! my poor Giovanni!" answered the Professor, with a calm expression of pity, "I know this wretched girl far better than yourself. You shall hear the truth in respect to the poisoner Rappaccini and his poisonous daughter; yes, poisonous as she is beautiful. Listen; for, even should you do violence to my gray hairs, it shall not silence me. That old fable of the Indian woman has become a truth by the deep and deadly science of Rappaccini and in the person of the lovely Beatrice."

Giovanni groaned and hid his face.

"Her father," continued Baglioni, "was not restrained by natural affection from offering up his child in this horrible manner as the victim of his insane zeal for science; for, let us do him justice, he is as true a man of science as ever distilled his own heart in an alembic. What, then, will be your fate? Beyond a doubt you are selected as the material of some new experiment. Perhaps the result is to be death; perhaps a fate more awful still. Rappaccini, with what he calls the interest of science before his eyes, will hesitate at nothing."

"It is a dream," muttered Giovanni to himself; "surely it is a dream."

"But," resumed the Professor, "be of good cheer, son of my friend. It is not yet too late for the rescue. Possibly we may even succeed in bringing back this miserable child within the limits of ordinary nature, from which her father's madness has estranged her. Behold this little silver vase! It was wrought by the hands of the renowned Benvenuto Cellini, and is well worthy to be a love gift to the fairest dame in Italy. But its contents are invaluable. One little sip of this antidote would have rendered the most virulent poisons of the Borgias innocuous. Doubt not that it will be as efficacious against those of Rappaccini. Bestow the vase, and the precious liquid within it, on your Beatrice, and hopefully await the result."

Baglioni laid a small, exquisitely wrought silver phial on the table and withdrew, leaving what he had said to produce its effect upon the young man's mind.

"We will thwart Rappaccini yet," thought he, chuckling to himself, as he descended the stairs; "but, let us confess the truth of him, he is a wonderful man—a wonderful man indeed; a vile empiric, however, in his practice, and therefore not to be tolerated by those who respect the good old rules of the medical profession."

Throughout Giovanni's whole acquaintance with Beatrice, he had occasionally, as we have said, been haunted by dark surmises as to her character; yet so thoroughly had she made herself felt by him as a simple, natural, most affectionate, and guileless creature, that the image now held up by Professor Baglioni looked as strange and incredible as if it were not in accordance with his own original conception. True, there were ugly recollections connected with his first glimpses of the beautiful girl; he could not quite forget the bouquet that withered in her grasp, and the insect that perished amid the sunny air, by no ostensible agency save the fragrance of her breath. These incidents, however, dissolving in the pure light of her character, had no longer the efficacy of facts, but were acknowledged as mistaken fantasies, by whatever testimony of the senses they might appear to be substantiated. There is something truer and more real than what we can see with the eyes and touch with the finger. On such better evidence had Giovanni founded his confidence in Beatrice, though rather by the necessary force of her high attributes than by any deep and generous faith on his part. But now his spirit was incapable of sustaining itself at the height to which the early enthusiasm of passion had exalted it; he fell down, grovelling among earthly doubts, and defiled therewith the pure whiteness of Beatrice's image. Not that he gave her up; he did but distrust. He resolved to institute some decisive test that should satisfy him, once for all, whether there were those dreadful peculiarities in her physical nature which could not be supposed to exist without some corresponding monstrosity of soul. His eyes, gazing down afar, might have deceived him as to the lizard, the insect, and the flowers; but if he could witness, at the distance of a few paces, the sudden blight of one fresh and healthful flower in Beatrice's hand, there would be room for no further question. With this idea, he hastened to the florist's and purchased a bouquet that was still gemmed with the morning dew-drops.

It was now the customary hour of his daily interview with Beatrice. Before descending into the garden, Giovanni failed not to look at his figure in the mirror,—a vanity to be expected in a beautiful young man, yet, as displaying itself at that troubled and feverish moment, the token of a certain shallowness of feeling and insincerity of character. He did gaze, however, and said to himself that his features had never before possessed so rich a grace, nor his eyes such vivacity, nor his cheeks so warm a hue of superabundant life.

"At least," thought he, "her poison has not yet insinuated itself into my system. I am no flower to perish in her grasp."

With that thought, he turned his eyes on the bouquet, which he had never once laid aside from his hand. A thrill of indefinable horror shot through his frame on perceiving that those dewy flowers were already beginning to droop; they wore the aspect of things that had been fresh and lovely yesterday. Giovanni grew white as marble, and stood motionless before the mirror, staring at his own reflection there as at the likeness of something frightful. He remembered Baglioni's remark about the fragrance that seemed to pervade the chamber. It must have been the poison in his breath! Then he shuddered—shuddered at himself. Recovering from his stupor, he began to watch with curious eye a spider that was busily at work hanging its web from the antique cornice of the apartment, crossing and recrossing the artful system of interwoven lines—as vigorous and active a spider as ever dangled from an old ceiling. Giovanni bent towards the insect, and emitted a deep, long breath. The spider suddenly ceased its toil; the web vibrated with a tremor originating in the body of the small artisan. Again Giovanni sent forth a breath, deeper, longer, and imbued with a venomous feeling out of his heart; he knew not whether he were wicked, or only desperate. The spider made a convulsive gripe with his limbs, and hung dead across the window.

"Accursed! accursed!" muttered Giovanni, addressing himself. "Hast thou grown so poisonous that this deadly insect perishes by thy breath?"

At that moment, a rich, sweet voice came floating up from the garden.

"Giovanni! Giovanni! It is past the hour! Why tarriest thou! Come down!"

"Yes," muttered Giovanni again. "She is the only being whom my breath may not slay! Would that it might!"

He rushed down, and in an instant was standing before the bright and loving eyes of Beatrice. A moment ago his wrath and despair had been so fierce that he could have desired nothing so much as to wither her by a glance; but with her actual presence there came influences which had too real an existence to be at once shaken off; recollections of the delicate and benign power of her feminine nature, which had so often enveloped him in a religious calm; recollections of many a holy and passionate outgush of her heart, when the pure fountain had been unsealed from its depths and made visible in its transparency to his mental eye; recollections which, had Giovanni known how to estimate them, would have assured him that all this ugly mystery was but an earthly illusion, and that, whatever mist of evil might seem to have gathered over her, the real Beatrice was a heavenly angel. Incapable as he was of such high faith, still her presence had not utterly lost its magic. Giovanni's rage was quelled into an aspect of sullen insensibility. Beatrice, with a quick spiritual sense, immediately felt that there was a gulf of blackness between them which neither he nor she could pass. They walked on together, sad and silent, and came thus to the marble fountain and to its pool of water on the ground, in the midst of which grew the shrub that bore gem-like blossoms. Giovanni was affrighted at the eager enjoyment—the appetite, as it were—with which he found himself inhaling the fragrance of the flowers.

"Beatrice," asked he abruptly, "whence came this shrub!"

"My father created it," answered she, with simplicity.

"Created it! created it!" repeated Giovanni. "What mean you, Beatrice?"

"He is a man fearfully acquainted with the secrets of Nature," replied Beatrice; "and, at the hour when I first drew breath, this plant sprang from the soil, the offspring of his science, of his intellect, while I was but his earthly child. Approach it not!" continued she, observing with terror that Giovanni was drawing nearer to the shrub. "It has qualities that you little dream of. But I, dearest Giovanni,—I grew up and blossomed with the plant and was nourished with its breath. It was my sister, and I loved it with a human affection; for—alas!—hast thou not suspected it?—there was an awful doom."

Here Giovanni frowned so darkly upon her that Beatrice paused and trembled. But her faith in his tenderness reassured her, and made her blush that she had doubted for an instant.

"There was an awful doom," she continued, "the effect of my father's fatal love of science, which estranged me from all society of my kind. Until Heaven sent thee, dearest Giovanni, Oh! how lonely was thy poor Beatrice!"

"Was it a hard doom?" asked Giovanni, fixing his eyes upon her.

"Only of late have I known how hard it was," answered she, tenderly. "Oh, yes; but my heart was torpid, and therefore quiet."

Giovanni's rage broke forth from his sullen gloom like a lightning flash out of a dark cloud.

"Accursed one!" cried he, with venomous scorn and anger. "And finding thy solitude wearisome, thou hast severed me, likewise, from all the warmth of life, and enticed me into thy region of unspeakable horror!"

"Giovanni!" exclaimed Beatrice, turning her large bright eyes upon his face. The force of his words had not found its way into her mind; she was merely thunderstruck.

"Yes, poisonous thing!" repeated Giovanni, beside himself with passion. "Thou hast done it! Thou hast blasted me! Thou hast filled my veins with poison! Thou hast made me as hateful, as ugly, as loathsome and deadly a creature as thyself—a world's wonder of hideous monstrosity! Now, if our breath be happily as fatal to ourselves as to all others, let us join our lips in one kiss of unutterable hatred, and so die!"

"What has befallen me?" murmured Beatrice, with a low moan out of her heart. "Holy Virgin pity me, a poor heart-broken child!"

"Thou,—dost thou pray?" cried Giovanni, still with the same fiendish scorn. "Thy very prayers, as they come from thy lips, taint the atmosphere with death. Yes, yes; let us pray! Let us to church, and dip our fingers in the holy water at the portal! They that come after us will perish as by a pestilence! Let us sign crosses in the air! It will be scattering curses abroad in the likeness of holy symbols!"

"Giovanni," said Beatrice, calmly, for her grief was beyond passion, "why dost thou join thyself with me thus in those terrible words? I, it is true, am the horrible thing thou namest me. But thou,—what hast thou to do, save with one other shudder at my hideous misery, to go forth out of the garden and mingle with thy race, and forget that there ever crawled on earth such a monster as poor Beatrice?"

"Dost thou pretend ignorance?" asked Giovanni, scowling upon her. "Behold! this power have I gained from the pure daughter of Rappaccini!"

There was a swarm of summer insects flitting through the air in search of the food promised by the flower odors of the fatal garden. They circled round Giovanni's head, and were evidently attracted towards him by the same influence which had drawn them for an instant within the sphere of several of the shrubs. He sent forth a breath among them, and smiled bitterly at Beatrice as at least a score of the insects fell dead upon the ground.

"I see it! I see it!" shrieked Beatrice. "It is my father's fatal science! No, no, Giovanni; it was not I! Never, never! I dreamed only to love thee, and be with thee a little time, and so to let thee pass away, leaving but thine image in mine heart; for, Giovanni, believe it, though my body be nourished with poison, my spirit is God's creature, and craves love as its daily food. But my father,—he has united us in this fearful sympathy. Yes; spurn me, tread upon me, kill me! Oh, what is death, after such words as thine? But it was not I. Not for a world of bliss would I have done it."

Giovanni's passion had exhausted itself in its outburst from his lips. There now came across him a sense, mournful, and not without tenderness, of the intimate and peculiar relationship between Beatrice and himself. They stood, as it were, in an utter solitude, which would be made none the less solitary by the densest throng of human life. Ought not, then, the desert of humanity around them to press this insulated pair closer together? If they should be cruel to one another, who was there to be kind to them? Besides, thought Giovanni, might there not still be a hope of his returning within the limits of ordinary nature, and leading Beatrice, the redeemed Beatrice, by the hand? O, weak, and selfish, and unworthy spirit, that could dream of an earthly union and earthly happiness as possible, after such deep love had been so bitterly wronged as was Beatrice's love by Giovanni's blighting words! No, no; there could be no such hope. She must pass heavily, with that broken heart, across the borders of Time—she must bathe her hurts in some fount of paradise, and forget her grief in the light of immortality, and *there* be well!

But Giovanni did not know it.

"Dear Beatrice," said he, approaching her, while she shrank away as always at his approach, but now with a different impulse, "dearest Beatrice, our fate is not yet so desperate. Behold! There is a medicine, potent, as a wise physician has assured me, and almost divine in its efficacy. It is composed of ingredients the most opposite to those by which thy awful father has brought this calamity upon thee and me. It is distilled of blessed herbs. Shall we not quaff it together, and thus be purified from evil?"

"Give it me!" said Beatrice, extending her hand to receive the little silver vial which Giovanni took from his bosom. She added, with a peculiar emphasis, "I will drink; but do thou await the result."

She put Baglioni's antidote to her lips; and, at the same moment, the figure of Rappaccini emerged from the portal and came slowly towards the marble fountain. As he drew near, the pale man of science seemed to gaze with a triumphant expression at the beautiful youth and maiden, as might an artist who should spend his life in achieving a picture or a group of statuary, and finally be satisfied with his success. He paused; his bent form grew erect with conscious power; he spread out his hand over them in the attitude of a father imploring a blessing upon his children; but those were the same hands that had thrown poison into the stream of their lives. Giovanni trembled. Beatrice shuddered very nervously, and pressed her hand upon her heart.

"My daughter," said Rappaccini, "thou art no longer lonely in the world! Pluck one of those precious gems from thy sister shrub, and bid thy bridegroom wear it in his bosom. It will not harm him now. My science and the sympathy between thee and him have so wrought within his system that he now stands apart from common men, as thou dost, daughter of my pride and triumph, from ordinary women. Pass on, then, through the world, most dear to one another and dreadful to all besides!"

"My father," said Beatrice, feebly,—and still as she spoke she kept her hand upon her heart—"wherefore didst thou inflict this miserable doom upon thy child?"

"Miserable!" exclaimed Rappaccini. "What mean you, foolish girl? Dost thou deem it misery to be endowed with marvellous gifts against which no power nor strength could avail an enemy—misery, to be able to quell the mightiest with a breath—misery, to be as terrible as thou art beautiful? Wouldst thou, then, have preferred the condition of a weak woman, exposed to all evil and capable of none?"

"I would fain have been loved, not feared," murmured Beatrice, sinking down upon the ground. "But now it matters not. I am going, father, where the evil which thou hast striven to mingle with my being will pass away like a dream—like the fragrance of these poisonous flowers, which will no longer taint my breath among the flowers of Eden. Farewell, Giovanni! Thy words of hatred are like lead within my heart; but they, too, will fall away as I ascend. Oh, was there not, from the first, more poison in thy nature than in mine?"

To Beatrice,—so radically had her earthly part been wrought upon by Rappaccini's skill,—as poison had been life, so the powerful antidote was death; and thus the poor victim of man's ingenuity and of thwarted nature, and of the fatality that attends all such efforts of perverted wisdom, perished there, at the feet of her father and Giovanni. Just at that moment Professor Pietro Baglioni looked forth from the window, and called loudly, in a tone of triumph mixed with horror, to the thunderstricken man of science,—

"Rappaccini! Rappaccini! and is *this* the upshot of your experiment?"

DISCUSSION QUESTIONS

1. Given that Rappaccini's daughter has a name, how does the title take on significance? In what ways does it establish themes of power and possession that inform the tale? In the preface, Hawthorne playfully identifies the author of the tale as an Italian writer named l'Aubépine, the French version of "Hawthorn," and the title of this story as "Beatrice; ou la Belle Empoisonneuse," which translates as "Beatrice: or, the Beautiful Poisoner." In what ways does this alternate title affect issues of agency and autonomy in the story? Consult Herndl (1993) for a comparison of these titles.

2. Rappaccini and Baglioni construct narratives about Beatrice that influence her fate. What are those narratives, and what do they also convey about the teller? How does each doctor's perspective reflect different attitudes toward the practice of medicine or the acquisition of knowledge through biomedical science?

3. What attempts does Beatrice make to control her own life and the narratives told about her? To what extent is she successful? In health care today, who has the authority to tell the patient's story? Make a list of health care providers who play some role in constructing the narrative of a patient's illness. Why is it important to consider the biases, assumptions, and other factors that might influence what or how the details of a patient's health care narrative are recorded?

4. Baglioni claims that Rappaccini has made Giovanni part of his experiment. Gather evidence from the story to support his claim. Also, consider other tales in this collection that include experimentation with human subjects: Hawthorne's "The Birthmark" and "Dr. Heidegger's Experiment" and Poe's "The Facts in the Case of M. Valdemar," for example. What might these tales convey about the complexities of informed consent in biomedical research?

5. Giovanni calls Beatrice "a monster" and says she is both "beautiful and terrible." Does science ever create "monsters" in the name of progress? If so, what are the ethical quandaries that surround this type of creation? Provide examples to support your claims. Also, can you imagine that a health care provider or biomedical researcher might describe his or her professional work as both "beautiful and terrible"? Explain your answer.

Ethan Brand

A Chapter from an Abortive Romance

January 1850, *Boston Weekly Museum,* as "The Unpardonable Sin.
From an Unpublished Work."

Nearly twenty years before Ethan Brand diagnoses himself as the bearer of the Unpardonable Sin, the village doctor diagnosed him a madman. The contrast between this and Brand's self-diagnosis calls us to question definitions of and diagnostic criteria for madness. Like the doctor, Ethan Brand was interested in the human mind and body, but unlike the doctor, he allowed his pursuit of knowledge to separate him from the community. Through Ethan Brand, Hawthorne criticizes the nineteenth-century enthusiasm for intellectualism and objectivism that allowed no room for human emotion. For Hawthorne, the extremes of intellectualization, isolation, and aspiration led to a degeneration of the spirit that ultimately spelled disaster for the intellectual/scientist/doctor and those on whom he experimented. Hawthorne's distrust of science is a romantic notion, a backlash against the age of reason, but his disdain for the overintellectualization and individualism also reflects his criticism of his transcendentalist peers (DeBakey 1968).

More often than not, Hawthorne can be seen rejecting any existing philosophy and staking out his own turf in a territory between various worldviews of mental health—the early-century assessment of psychology as a "science of the soul," the evolving theory that mental illness was rooted in physical illness, and the late-century view of psychology as a science of the mind (Thraillkill 2006; Goldman 2004). Hawthorne rejected the bifurcation of mental illness into "moral" and "physical" categories and through his writing seems to intuit the need for a more holistic approach to healing, where even the artist has a role.

Certainly, as a writer Hawthorne played a role in problematizing the easy categorization of illness, suggesting in stories like "Ethan Brand" that observation alone cannot be equated with fact, as any action, body, or symbol may have multiple meanings and, thus, require interpretation (Browner 1993). In tales like this one and "Egotism; or, the Bosom Serpent," Hawthorne provides compelling evidence for the romantic notion that writers as artists can broaden our understanding of the illness experience through story.

•

BARTRAM the lime-burner, a rough, heavy-looking man, begrimed with charcoal, sat watching his kiln at nightfall, while his little son played at building houses with the scattered fragments of marble, when, on the hill-side below them, they heard a roar of laughter, not mirthful, but slow, and even solemn, like a wind shaking the boughs of the forest.

"Father, what is that?" asked the little boy, leaving his play, and pressing betwixt his father's knees.

"Oh, some drunken man, I suppose," answered the lime-burner; "some merry fellow from the bar-room in the village, who dared not laugh loud enough within doors lest he should blow the roof of the house off. So here he is, shaking his jolly sides at the foot of Graylock."

"But, father," said the child, more sensitive than the obtuse, middle-aged clown, "he does not laugh like a man that is glad. So the noise frightens me!"

"Don't be a fool, child!" cried his father, gruffly. "You will never make a man, I do believe; there is too much of your mother in you. I have known the rustling of a leaf startle you. Hark! Here comes the merry fellow now. You shall see that there is no harm in him."

Bartram and his little son, while they were talking thus, sat watching the same lime-kiln that had been the scene of Ethan Brand's solitary and meditative life, before he began his search for the Unpardonable Sin. Many years, as we have seen, had now elapsed, since that portentous night when the IDEA was first developed. The kiln, however, on the mountain-side, stood unimpaired, and was in nothing changed since he had thrown his dark thoughts into the intense glow of its furnace, and melted them, as it were, into the one thought that took possession of his life. It was a rude, round, tower-like structure about twenty feet high, heavily built of rough stones, and with a hillock of earth heaped about the larger part of its circumference; so that the blocks and fragments of marble might be drawn by cart-loads, and thrown in at the top. There was an opening at the bottom of the tower, like an oven-mouth, but large enough to admit a man in a stooping posture, and provided with a massive iron door. With the smoke and jets of flame issuing from the chinks and crevices of this door, which seemed to give admittance into the hill-side, it resembled nothing so much as the private entrance to the infernal regions, which the shepherds of the Delectable Mountains were accustomed to show to pilgrims.

There are many such lime-kilns in that tract of country, for the purpose of burning the white marble which composes a large part of the substance of the hills. Some of them, built years ago, and long deserted, with weeds growing in the vacant round of the interior, which is open to the sky, and grass and wild-flowers rooting themselves into the chinks of the stones, look already like relics of antiquity, and may yet be overspread with the lichens of centuries to come. Others, where the lime-burner still feeds his daily and nightlong fire, afford points of interest to the wanderer among the hills, who seats himself on a log of wood or a fragment of marble, to hold a chat with the solitary man. It is a lonesome, and, when the character is inclined to thought, may be an intensely thoughtful occupation; as it proved in the case of Ethan Brand, who had mused to such strange purpose, in days gone by, while the fire in this very kiln was burning.

The man who now watched the fire was of a different order, and troubled himself with no thoughts save the very few that were requisite to his business. At frequent intervals, he flung back the clashing weight of the iron door, and, turning his face from the insufferable glare, thrust in huge logs of oak, or stirred the immense brands with a long pole. Within the furnace were seen the curling and riotous flames, and the burning marble, almost molten with the intensity of heat; while without, the reflection of the fire quivered on the dark intricacy of the surrounding forest, and showed in the foreground a bright and ruddy little picture of the hut, the spring beside its door, the athletic and coal-begrimed figure of the lime-burner, and the half-frightened child, shrinking into the protection of his father's shadow. And when again the iron door was closed, then reappeared the tender light of the half-full moon, which vainly strove to trace out the indistinct shapes of the neighboring mountains; and, in the upper sky, there was a flitting congregation of clouds, still faintly tinged with the rosy sunset, though thus far down into the valley the sunshine had vanished long and long ago.

The little boy now crept still closer to his father, as footsteps were heard ascending the hill-side, and a human form thrust aside the bushes that clustered beneath the trees.

"Halloo! who is it?" cried the lime-burner, vexed at his son's timidity, yet half infected by it. "Come forward, and show yourself, like a man, or I'll fling this chunk of marble at your head!"

"You offer me a rough welcome," said a gloomy voice, as the unknown man drew nigh. "Yet I neither claim nor desire a kinder one, even at my own fireside."

To obtain a distincter view, Bartram threw open the iron door of the kiln, whence immediately issued a gush of fierce light, that smote full upon the stranger's face and figure. To a careless eye there appeared nothing very remark-able in his aspect, which was that of a man in a coarse, brown, country-made suit of clothes, tall and thin, with the staff and heavy shoes of a wayfarer. As he advanced, he fixed his eyes—which were very bright—intently upon the brightness of the furnace, as if he beheld, or expected to behold, some object worthy of note within it.

"Good evening, stranger," said the lime-burner; "whence come you, so late in the day?"

"I come from my search," answered the wayfarer; "for, at last, it is finished."

"Drunk!—or crazy!" muttered Bartram to himself. "I shall have trouble with the fellow. The sooner I drive him away, the better."

The little boy, all in a tremble, whispered to his father, and begged him to shut the door of the kiln, so that there might not be so much light; for that there was something in the man's face which he was afraid to look at, yet could not look away from. And, indeed, even the lime-burner's dull and torpid sense began to be impressed by an indescribable something in that thin, rugged, thoughtful visage, with the grizzled hair hanging wildly about it, and those deeply sunken eyes, which gleamed like fires within the entrance of a mysterious cavern. But, as he closed the door, the stranger turned towards him, and spoke in a quiet, familiar way, that made Bartram feel as if he were a sane and sensible man, after all.

"Your task draws to an end, I see," said he. "This marble has already been burning three days. A few hours more will convert the stone to lime."

"Why, who are you?" exclaimed the lime-burner. "You seem as well acquainted with my business as I am myself."

"And well I may be," said the stranger; "for I followed the same craft many a long year, and here, too, on this very spot. But you are a newcomer in these parts. Did you never hear of Ethan Brand?"

"The man that went in search of the Unpardonable Sin?" asked Bartram, with a laugh.

"The same," answered the stranger. "He has found what he sought, and therefore he comes back again."

"What! then you are Ethan Brand himself?" cried the lime-burner, in amazement. "I am a newcomer here, as you say, and they call it eighteen years since you left the foot of Graylock. But, I can tell you, the good folks still talk about Ethan Brand, in the village yonder, and what a strange errand took him away from his lime-kiln. Well, and so you have found the Unpardonable Sin?"

"Even so!" said the stranger, calmly.

"If the question is a fair one," proceeded Bartram, "where might it be?"

Ethan Brand laid his finger on his own heart.

"Here!" replied he.

And then, without mirth in his countenance, but as if moved by an involuntary recognition of the infinite absurdity of seeking throughout the world for what was the closest of all things to himself, and looking into every heart, save his own, for what was hidden in no other breast, he broke into a laugh of scorn. It was the same slow, heavy laugh, that had almost appalled the lime-burner when it heralded the wayfarer's approach.

The solitary mountain-side was made dismal by it. Laughter, when out of place, mistimed, or bursting forth from a disordered state of feeling, may be the most terrible modulation of the human voice. The laughter of one asleep, even if it be a little child,—the madman's laugh,—the wild, screaming laugh of a born idiot,—are sounds that we sometimes tremble to hear, and would always willingly forget. Poets have imagined no utterance of fiends or hobgoblins so fearfully appropriate as a laugh. And even the obtuse lime-burner felt his nerves shaken, as this strange man looked inward at his own heart, and burst into laughter that rolled away into the night, and was indistinctly reverberated among the hills.

"Joe," said he to his little son," scamper down to the tavern in the village, and tell the jolly fellows there that Ethan Brand has come back, and that he has found the Unpardonable Sin!"

The boy darted away on his errand, to which Ethan Brand made no objection, nor seemed hardly to notice it. He sat on a log of wood, looking steadfastly at the iron door of the kiln. When the child was out of sight, and his swift and light footsteps ceased to be heard treading first on the fallen leaves and then on the rocky mountain path, the lime-burner began to regret his departure. He felt that the little fellow's presence had been a barrier between his guest and himself, and that he must now deal, heart to heart, with a man who, on his own confession, had committed the one only crime for which Heaven could afford no mercy. That crime, in its indistinct blackness, seemed to overshadow him. The lime-burner's own sins rose up within him, and made his memory riotous with a throng of evil shapes that asserted their kindred with the Master Sin, whatever it might be, which it was within the scope of man's corrupted nature to conceive and cherish. They were all of one family; they went to and fro between his breast and Ethan Brand's, and carried dark greetings from one to the other.

Then Bartram remembered the stories which had grown traditionary in reference to this strange man, who had come upon him like a shadow of the night, and was making himself at home in his old place, after so long absence that the dead people, dead and buried for years, would have had more right to be at home, in any familiar spot, than he. Ethan Brand, it was said, had conversed with Satan himself in the lurid blaze of this very kiln. The legend had been matter of mirth heretofore, but looked grisly now. According to this tale, before Ethan Brand departed on his search, he had been accustomed to evoke a fiend from the hot furnace of the lime-kiln, night after night, in order to confer with him about the Unpardonable Sin; the man and the fiend each laboring to frame the image of some mode of guilt which could neither be atoned for nor forgiven. And, with the first gleam of light upon the mountain-top, the fiend crept in at the iron door, there to abide the intensest element of fire, until again summoned forth to share in the dreadful task of extending man's possible guilt beyond the scope of Heaven's else infinite mercy.

While the lime-burner was struggling with the horror of these thoughts, Ethan Brand rose from the log, and flung open the door of the kiln. The action was in such accordance with the idea in Bartram's mind, that he almost expected to see the Evil One issue forth, red-hot, from the raging furnace.

"Hold! hold!" cried he, with a tremulous attempt to laugh; for he was ashamed of his fears, although they overmastered him. "Don't, for mercy's sake, bring out your Devil now!"

"Man!" sternly replied Ethan Brand, "what need have I of the Devil? I have left him behind me, on my track. It is with such half-way sinners as you that he busies himself. Fear not, because I open the door. I do but act by old custom, and am going to trim your fire, like a lime-burner, as I was once."

He stirred the vast coals, thrust in more wood, and bent forward to gaze into the hollow prison-house of the fire, regardless of the fierce glow that reddened upon his face. The lime-burner sat watching him, and half suspected his strange guest of a purpose, if not to evoke a fiend, at least to plunge bodily into the flames, and thus vanish from the sight of man. Ethan Brand, however, drew quietly back, and closed the door of the kiln.

"I have looked," said he, "into many a human heart that was seven times hotter with sinful passions than yonder furnace is with fire. But I found not there what I sought. No, not the Unpardonable Sin!"

"What is the Unpardonable Sin?" asked the lime-burner; and then he shrank further from his companion, trembling lest his question should be answered.

"It is a sin that grew within my own breast," replied Ethan Brand, standing erect, with a pride that distinguishes all enthusiasts of his stamp. "A sin that grew nowhere else! The sin of an intellect that triumphed over the sense of brotherhood with man and reverence for God, and sacrificed everything to its own mighty claims! The only sin that deserves a recompense of immortal agony! Freely, were it to do again, would I incur the guilt. Unshrinkingly I accept the retribution!"

"The man's head is turned," muttered the lime-burner to himself. "He may be a sinner, like the rest of us,—nothing more likely,—but, I'll be sworn, he is a madman too."

Nevertheless he felt uncomfortable at his situation, alone with Ethan Brand on the wild mountain-side, and was right glad to hear the rough murmur of tongues, and the footsteps of what seemed a pretty numerous party, stumbling over the stones and rustling through the underbrush. Soon appeared the whole lazy regiment that was wont to infest the village tavern, comprehending three or four individuals who had drunk flip beside the bar-room fire through all the winters, and smoked their pipes beneath the stoop through all the summers, since Ethan Brand's departure. Laughing boisterously, and mingling all their voices together in unceremonious talk, they now burst into the moonshine and narrow streaks of firelight that illuminated the open space before the lime-kiln. Bartram set the door ajar again, flooding the spot with light, that the whole company might get a fair view of Ethan Brand, and he of them.

There, among other old acquaintances, was a once ubiquitous man, now almost extinct, but whom we were formerly sure to encounter at the hotel of every thriving village throughout the country. It was the stage-agent. The present specimen of the genus was a wilted and smoke-dried man, wrinkled and red-nosed, in a smartly cut, brown, bobtailed coat, with brass buttons, who, for a length of time unknown, had kept his desk and corner in the bar-room, and was still puffing what seemed to be the same cigar that he had lighted twenty years before. He had great fame as a dry joker, though, perhaps, less on account of any intrinsic humor than from a certain flavor of brandy-toddy and tobacco-smoke, which impregnated all his ideas and expressions, as well as his person. Another well-remembered, though strangely altered, face was that of Lawyer Giles, as people still called him in courtesy; an elderly ragamuffin, in his soiled shirt-sleeves and tow-cloth trousers. This poor fellow had been an attorney, in what he called his better days, a sharp practitioner, and in great vogue among the village litigants; but flip, and sling, and toddy, and cocktails, imbibed at all hours, morning, noon, and night, had caused him to slide from intellectual to various kinds and degrees of bodily labor, till, at last, to adopt his own phrase, he slid into a soap-vat. In other words, Giles was now a soap-boiler, in a small way. He had come to be but the fragment of a human being, a part of one foot having been chopped off by an axe, and an entire hand torn away by the devilish grip of a steam-engine. Yet, though the corporeal hand was gone, a spiritual member remained; for, stretching forth the stump, Giles steadfastly averred that he felt an invisible thumb and fingers with as vivid a sensation as before the real ones were amputated. A maimed and miserable wretch he was; but one, nevertheless, whom the world could not trample on, and had no right to scorn, either in this or any previous stage of his misfortunes, since he had still kept up the courage and spirit of a man, asked nothing in charity, and with his one hand—and that the left one—fought a stern battle against want and hostile circumstances.

Among the throng, too, came another personage, who, with certain points of similarity to Lawyer Giles, had many more of difference. It was the village doctor; a man of some fifty years, whom, at an earlier period of his life, we introduced as paying a professional visit to Ethan Brand during the latter's supposed insanity. He was now a purple-visaged, rude, and brutal, yet half-gentlemanly figure, with something wild, ruined, and desperate in his talk, and in all the details of his gesture and manners. Brandy possessed this man like an evil spirit, and made him as surly and savage as a wild beast, and as miserable as a lost soul; but there was supposed to be in him such wonderful skill, such native gifts of healing, beyond any which medical science could impart, that society caught hold of him, and would not let him sink out of its reach. So, swaying to and fro upon his horse, and grumbling thick accents at the bedside, he visited all the sick chambers for miles about among the mountain towns, and sometimes raised a dying man, as it were, by miracle, or quite as often, no doubt, sent his patient to a grave that was dug many a year too soon. The doctor had an everlasting pipe in his mouth, and, as somebody said, in allusion to his habit of swearing, it was always alight with hell-fire.

These three worthies pressed forward, and greeted Ethan Brand each after his own fashion, earnestly inviting him to partake of the contents of a certain black bottle, in which, as they averred, he would find something far better worth seeking for than the Unpardonable Sin. No mind, which has wrought itself by intense and solitary meditation into a high state of enthusiasm, can endure the kind of contact with low and vulgar modes of thought and feeling to which Ethan Brand was now subjected. It made him doubt—and, strange to say, it was a painful doubt—whether he had indeed found the Unpardonable Sin, and found it within himself. The whole question on which he had exhausted life, and more than life, looked like a delusion.

"Leave me," he said, bitterly, "ye brute beasts, that have made yourselves so, shrivelling up your souls with fiery liquors! I have done with you. Years and years ago, I groped into your hearts and found nothing there for my purpose. Get ye gone!"

"Why, you uncivil scoundrel," cried the fierce doctor, "is that the way you respond to the kindness of your best friends? Then let me tell you the truth. You have no more found the Unpardonable Sin than yonder boy Joe has. You are but a crazy fellow,—I told you so twenty years ago,—neither better nor worse than a crazy fellow, and the fit companion of old Humphrey, here!"

He pointed to an old man, shabbily dressed, with long white hair, thin visage, and unsteady eyes. For some years past this aged person had been wandering about among the hills, inquiring of all travellers whom he met for his daughter. The girl, it seemed, had gone off with a company of circus-performers, and occasionally tidings of her came to the village, and fine stories were told of her glittering appearance as she rode on horseback in the ring, or performed marvellous feats on the tight-rope.

The white-haired father now approached Ethan Brand, and gazed unsteadily into his face.

"They tell me you have been all over the earth," said he, wringing his hands with earnestness. "You must have seen my daughter, for she makes a grand figure in the world, and everybody goes to see her. Did she send any word to her old father, or say when she was coming back?"

Ethan Brand's eye quailed beneath the old man's. That daughter, from whom he so earnestly desired a word of greeting, was the Esther of our tale, the very girl whom, with such cold and remorseless purpose, Ethan Brand had made the subject of a psychological experiment, and wasted, absorbed, and perhaps annihilated her soul, in the process.

"Yes," murmured he, turning away from the hoary wanderer, "it is no delusion. There is an Unpardonable Sin!"

While these things were passing, a merry scene was going forward in the area of cheerful light, beside the spring and before the door of the hut. A number of the youth of the village, young men and girls, had hurried up the hill-side, impelled by curiosity to see Ethan Brand, the hero of so many a legend familiar to their childhood. Finding nothing, however, very remarkable in his aspect,—nothing but a sunburnt wayfarer, in plain garb and dusty shoes, who sat looking into the fire, as if he fancied pictures among the coals,—these young people speedily grew tired of observing him. As it happened, there was other amusement at hand. An old German Jew, travelling with a diorama on his back, was passing down the mountain-road towards the village just as the party turned aside from it, and, in hopes of eking out the profits of the day, the showman had kept them company to the lime-kiln.

"Come, old Dutchman," cried one of the young men, "let us see your pictures, if you can swear they are worth looking at!"

"O, yes, Captain," answered the Jew,—whether as a matter of courtesy or craft, he styled everybody Captain,—"I shall show you, indeed, some very superb pictures!"

So, placing his box in a proper position, he invited the young men and girls to look through the glass orifices of the machine, and proceeded to exhibit a series of the most outrageous scratchings and daubings, as specimens of the fine arts, that ever an itinerant showman had the face to impose upon his circle of spectators. The pictures were worn out, moreover, tattered, full of cracks and wrinkles, dingy with tobacco-smoke, and otherwise in a most pitiable condition. Some purported to be cities, public edifices, and ruined castles in Europe; others represented Napoleon's battles and Nelson's sea-fights; and in the midst of these would be seen a gigantic, brown, hairy hand,—which might have been mistaken for the Hand of Destiny, though, in truth, it was only the showman's,—pointing its forefinger to various scenes of the conflict, while its owner gave historical illustrations. When, with much merriment at its abominable deficiency of merit, the exhibition was concluded, the German bade little Joe put his head into the box. Viewed through the magnifying-glasses, the boy's round, rosy visage assumed the strangest imaginable aspect of an immense Titanic child, the mouth grinning broadly, and the eyes and every other feature overflowing with fun at the joke. Suddenly, however, that merry face turned pale, and its expression changed to horror, for this easily impressed and excitable child had become sensible that the eye of Ethan Brand was fixed upon him through the glass.

"You make the little man to be afraid, Captain," said the German Jew, turning up the dark and strong outline of his visage, from his stooping posture. "But look again, and, by chance, I shall cause you to see somewhat that is very fine, upon my word!"

Ethan Brand gazed into the box for an instant, and then starting back, looked fixedly at the German. What had he seen? Nothing, apparently; for a curious youth, who had peeped in almost at the same moment, beheld only a vacant space of canvas.

"I remember you now," muttered Ethan Brand to the showman.

"Ah, Captain," whispered the Jew of Nuremberg, with a dark smile, "I find it to be a heavy matter in my show-box—this Unpardonable Sin! By my faith, Captain, it has wearied my shoulders, this long day, to carry it over the mountain."

"Peace," answered Ethan Brand, sternly, "or get thee into the furnace yonder!"

The Jew's exhibition had scarcely concluded, when a great, elderly dog—who seemed to be his own master, as no person in the company laid claim to him—saw fit to render himself the object of public notice. Hitherto, he had shown himself a very quiet, well-disposed old dog, going round from one to another, and, by way of being sociable, offering his rough head to be patted by any kindly hand that would take so much trouble. But now, all of a sudden, this grave and venerable quadruped, of his own mere motion, and without the slightest suggestion from anybody else, began to run round after his tail, which, to heighten the absurdity of the proceeding, was a great deal shorter than it should have been. Never was seen such headlong eagerness in pursuit of an object that could not possibly be attained; never was heard such a tremendous outbreak of growling, snarling, barking, and snapping,—as if one end of the ridiculous brute's body were at deadly and most unforgivable enmity with the other. Faster and faster, round about went the cur; and faster and still faster fled the unapproachable brevity of his tail; and louder and fiercer grew his yells of rage and animosity; until, utterly exhausted, and as far from the goal as ever, the foolish old dog ceased his performance as suddenly as he had begun it. The next moment he was as mild, quiet, sensible, and respectable in his deportment, as when he first scraped acquaintance with the company.

As may be supposed, the exhibition was greeted with universal laughter, clapping of hands, and shouts of encore, to which the canine performer responded by wagging all that there was to wag of his tail, but appeared totally unable to repeat his very successful effort to amuse the spectators.

Meanwhile, Ethan Brand had resumed his seat upon the log, and moved, it might be, by a perception of some remote analogy between his own case and that of this self-pursuing cur, he broke into the awful laugh, which, more than any other token, expressed the condition of his inward being. From that moment, the merriment of the party was at an end; they stood aghast, dreading lest the inauspicious sound should be reverberated around the horizon, and that mountain would thunder it to mountain, and so the horror be prolonged upon their ears. Then; whispering one to another that it was late,—that the moon was almost down,—that the August night was growing chill,—they hurried homewards, leaving the lime-burner and little Joe to deal as they might with their unwelcome guest. Save for these three human beings, the open space on the hill-side was a solitude, set in a vast gloom of forest. Beyond that darksome verge, the firelight glimmered on the stately trunks and almost black foliage of pines, intermixed with the lighter verdure of sapling oaks, maples, and poplars, while here and there lay the gigantic corpses of dead trees, decaying on the leaf-strewn soil. And it seemed to little Joe—a timorous and imaginative child—that the silent forest was holding its breath until some fearful thing should happen.

Ethan Brand thrust more wood into the fire, and closed the door of the kiln, then looking over his shoulder at the lime-burner and his son, he bade, rather than advised, them to retire to rest.

"For myself, I cannot sleep," said he. "I have matters that it concerns me to meditate upon. I will watch the fire, as I used to do in the old time."

"And call the Devil out of the furnace to keep you company, I suppose," muttered Bartram, who had been making intimate acquaintance with the black bottle above mentioned. "But watch, if you like, and call as many devils as you like! For my part, I shall be all the better for a snooze. Come, Joe!"

As the boy followed his father into the hut, he looked back at the wayfarer, and the tears came into his eyes, for his tender spirit had an intuition of the bleak and terrible loneliness in which this man had enveloped himself.

When they had gone, Ethan Brand sat listening to the crackling of the kindled wood, and looking at the little spirts of fire that issued through the chinks of the door. These trifles, however, once so familiar, had but the slightest hold of his attention, while deep within his mind he was reviewing the gradual but marvellous change that had been wrought upon him by the search to which he had devoted himself. He remembered how the night dew had fallen upon him,—how the dark forest had whispered to him,—how the stars had gleamed upon him,—a simple and loving man, watching his fire in the years gone by, and ever musing as it burned. He remembered with what tenderness, with what love and sympathy for mankind, and what pity for human guilt and woe, he had first begun to contemplate those ideas which afterwards became the inspiration of his life; with what reverence he had then looked into the heart of man, viewing it as a temple originally divine, and, however desecrated, still to be held sacred by a brother; with what awful fear he had deprecated the success of his pursuit, and prayed that the Unpardonable Sin might never be revealed to him. Then ensued that vast intellectual development, which, in its progress, disturbed the counterpoise between his mind and heart. The Idea that possessed his life had operated as a means of education; it had gone on cultivating his powers to the highest point of which they were susceptible; it had raised him from the level of an unlettered laborer to stand on a star-lit eminence, whither the philosophers of the earth, laden with the lore of universities, might vainly strive to clamber after him. So much for the intellect! But where was the heart? That, indeed, had withered,—had contracted,—had hardened,—had perished! It had ceased to partake of the universal throb. He had lost his hold of the magnetic chain of humanity. He was no longer a brother-man, opening the chambers or the dungeons of our common nature by the key of holy sympathy, which gave him a right to share in all its secrets; he was now a cold observer, looking on mankind as the subject of his experiment, and, at length, converting man and woman to be his puppets, and pulling the wires that moved them to such degrees of crime as were demanded for his study.

Thus Ethan Brand became a fiend. He began to be so from the moment that his moral nature had ceased to keep the pace of improvement with his intellect. And now, as his highest effort and inevitable development,—as the bright and gorgeous flower, and rich, delicious fruit of his life's labor,—he had produced the Unpardonable Sin!

"What more have I to seek? what more to achieve?" said Ethan Brand to himself. "My task is done, and well done!"

Starting from the log with a certain alacrity in his gait and ascending the hillock of earth that was raised against the stone circumference of the lime-kiln, he thus reached the top of the structure. It was a space of perhaps ten feet across, from edge to edge, presenting a view of the upper surface of the immense mass of broken marble with which the kiln was heaped. All these innumerable blocks and fragments of marble were red-hot and vividly on fire, sending up great spouts of blue flame, which quivered aloft and danced madly, as within a magic circle, and sank and rose again, with continual and multitudinous activity. As the lonely man bent forward over this terrible body of fire, the blasting heat smote up against his person with a breath that, it might be supposed, would have scorched and shrivelled him up in a moment.

Ethan Brand stood erect, and raised his arms on high. The blue flames played upon his face, and imparted the wild and ghastly light which alone could have suited its expression; it was that of a fiend on the verge of plunging into his gulf of intensest torment.

"O Mother Earth," cried he, "who art no more my Mother, and into whose bosom this frame shall never be resolved! O mankind, whose brotherhood I have cast off, and trampled thy great heart beneath my feet! O stars of heaven, that shone on me of old, as if to light me onward and upward!—farewell all, and forever. Come, deadly element of Fire,—henceforth my familiar friend! Embrace me, as I do thee!"

That night the sound of a fearful peal of laughter rolled heavily through the sleep of the lime-burner and his little son; dim shapes of horror and anguish haunted their dreams, and seemed still present in the rude hovel, when they opened their eyes to the daylight.

"Up, boy, up!" cried the lime-burner, staring about him. "Thank Heaven, the night is gone, at last; and rather than pass such another, I would watch my lime-kiln, wide awake, for a twelvemonth. This Ethan Brand, with his humbug of an Unpardonable Sin, has done me no such mighty favor, in taking my place!"

He issued from the hut, followed by little Joe, who kept fast hold of his father's hand. The early sunshine was already pouring its gold upon the mountain-tops, and though the valleys were still in shadow, they smiled cheerfully in the promise of the bright day that was hastening onward. The village, completely shut in by hills, which swelled away gently about it, looked as if it had rested peacefully in the hollow of the great hand of Providence. Every dwelling was distinctly visible; the little spires of the two churches pointed upwards, and caught a fore-glim-mering of brightness from the sun-gilt skies upon their gilded weather-cocks. The tavern was astir, and the figure of the old, smoke-dried stage-agent, cigar in mouth, was seen beneath the stoop. Old Graylock was glorified with a golden cloud upon his head. Scattered likewise over the breasts of the surrounding mountains, there were heaps of hoary mist, in fantastic shapes, some of them far down into the valley, others high up towards the summits, and still others, of the same family of mist or cloud, hovering in the gold radiance of the upper atmosphere. Stepping from one to another of the clouds that rested on the hills, and thence to the loftier brotherhood that sailed in air, it seemed almost as if a mortal man might thus ascend into the heavenly regions. Earth was so mingled with sky that it was a day-dream to look at it.

To supply that charm of the familiar and homely, which Nature so readily adopts into a scene like this, the stage-coach was rattling down the mountain-road, and the driver sounded his horn, while echo caught up the notes, and intertwined them into a rich and varied and elaborate harmony, of which the original performer could lay claim to little share. The great hills played a concert among themselves, each contributing a strain of airy sweetness.

Little Joe's face brightened at once.

"Dear father," cried he, skipping cheerily to and fro, "that strange man is gone, and the sky and the mountains all seem glad of it!"

"Yes," growled the lime-burner, with an oath, "but he has let the fire go down, and no thanks to him if five hundred bushels of lime are not spoiled. If I catch the fellow hereabouts again, I shall feel like tossing him into the furnace!"

With his long pole in his hand, he ascended to the top of the kiln. After a moment's pause, he called to his son.

"Come up here, Joe!" said he.

So little Joe ran up the hillock, and stood by his father's side. The marble was all burnt into perfect, snow-white lime. But on its surface, in the midst of the circle,—snow-white too, and thoroughly converted into lime,—lay a human skeleton, in the attitude of a person who, after long toil, lies down to long repose. Within the ribs—strange to say—was the shape of a human heart.

"Was the fellow's heart made of marble?" cried Bartram, in some perplexity at this phenomenon. "At any rate, it is burnt into what looks like special good lime; and, taking all the bones together, my kiln is half a bushel the richer for him."

So saying, the rude lime-burner lifted his pole, and, letting it fall upon the skeleton, the relics of Ethan Brand were crumbled into fragments.

DISCUSSION QUESTIONS

1. Nearly twenty years before Ethan Brand returns, claiming he has found the Unpardonable Sin, the village doctor diagnosed him as mad. Do you agree with this diagnosis? Why or why not?

2. Analyze the paragraph in which Hawthorne describes the village doctor. What are his strengths and weaknesses? How would you describe the qualities he possesses that have endeared him to the community? In what ways do community attitudes toward the health care system in the twenty-first century affect the care members of that community seek out—or resist seeking out?

3. What is the Unpardonable Sin, according to Ethan Brand? How does Brand's obsession with it reflect nineteenth-century anxieties about evolving trends in medicine? Do we experience comparable anxieties about health care today?

4. Ethan Brand believes that his "moral nature had ceased to keep the pace of improvement with his intellect." In health care practice or biomedical research, do patients suffer—directly or indirectly—when providers or researchers value intellect over "heart"? What measures can educators take to ensure that health care professions and biomedical research students develop both intellectual capacity and compassion during their educational years?

5. Characterize Bartram's response to Ethan Brand's suicide. What does his response say about his character? What is his interpretation of the heart shape he finds? What is your interpretation?

6. Lawyer Giles experiences a "phantom limb" approximately twenty years before Silas Weir Mitchell coined the term (1871). Some scholars and physicians have noted that Poe also described illnesses that the medico-scientific community had yet to understand fully or name. Based on these examples, consider the validity of the romantics' belief that both rational and imaginative approaches are valuable in understanding suffering and illness. Speculate on the role writers and artists can play in contributing to twenty-first century medico-scientific knowledge or, more directly, to healing patients.

7. Brand has chosen a type of death that ensures his body will not be available for medical observation. Does this seem a fitting act, given his reasons for dying? Today, medical educators still rely on the laboratory study of deceased bodies to teach anatomy. When a person donates his or her body to science, what measures are in place to ensure the donor's dignity is preserved before, during, and after study?

Edgar Allan Poe

Sonnet—To Science

1829, *Al Aaraaf, Tamerlane, and Minor Poems*

Long before the terms "medical" and "health humanities" emerged, Edgar Allan Poe espoused the romantic philosophy that science without art created a dangerous imbalance, and during his career he produced a number of works that promoted the marriage of rationality and creativity as an ideal method of inquiry. Notably, his famous sleuth, Dupin—America's first modern literary detective—excels because he combines his powers of observation and analysis with a keen intuition, giving him almost superhuman powers of deduction. But Dupin came much later. Early on in his career, Poe wrote "Sonnet—To Science," a poem expressing his criticism of scientific dogmatism that ignored human beings' need for poetry, myth, and creativity.

Poe believed art was another way of knowing and that the knowledge gained from humanistic pursuit complemented that gained from scientific inquiry (Levine and Levine 1990); in this way, Poe anticipates the health humanities. Today, Poe would likely agree with physician and literary scholar Rita Charon, who has proposed that narrative medicine and evidence-based medicine should not be dichotomized but combined to produce a more comprehensive base of evidence (Engel et al. 2008). Charon, along with numerous other providers, including physician-writer Abraham Verghese and nurse-writer Cortney Davis, identifies an increasing reliance on technology as interfering with the process of healing: patients are sent for more labs and tests and often receive less hands-on, therapeutic time with health care providers (Verghese 2011; Engel et al. 2008). Thus, Poe's themes remain relevant in twenty-first-century America, prompting us to consider how technological changes can both improve and detract from aspects of care, such as the physical exam, patient-provider relationships, and communication.

•

SONNET—TO SCIENCE

SCIENCE! true daughter of Old Time thou art!
> Who alterest all things with thy peering eyes.
Why preyest thou thus upon the poet's heart,
> Vulture, whose wings are dull realities?
How should he love thee? or how deem thee wise,
> Who wouldst not leave him in his wandering
To seek for treasure in the jewelled skies,
> Albeit he soared with an undaunted wing?
Hast thou not dragged Diana from her car?
> And driven the Hamadryad from the wood
To seek a shelter in some happier star?
> Hast thou not torn the Naiad from her flood,
The Elfin from the green grass, and from me
> The summer dream beneath the tamarind tree?

DISCUSSION QUESTIONS

1. The speaker of "Sonnet—To Science" is critical of Science, whom he addresses directly, using the poetic technique of apostrophe. What are the speaker's main criticisms? What motivates these criticisms? Do you agree or disagree? Why?

2. Using Poe's poem as a starting point, propose an ideal relationship between rationality and creativity. What role, for example, does or should creativity play in health care research and practice? Physicians sometimes refer to the practice of medicine as an "art"; how would the speaker of this poem respond to that description? Do other health care providers define their practice as an art? If so, how do they define the term?

3. Poe was a dark American romantic. Romantics responded to what they saw as the reckless promotion of technology and industry that grew out of the Enlightenment and corrupted man and society. Would you consider the field of health humanities a descendent of romanticism? Provide at least three reasons to support your answer.

4. The rise of clinical medicine in the nineteenth century placed emphasis on objective observation. What imagery in the poem seems to relate directly to a criticism of this reductive understanding of science? Would Poe offer similar criticisms about evidence-based medicine today? Would he view quantitative and qualitative research differently? Do you believe the two equally valuable and complementary? Explain your answer.

5. List other images that are particularly memorable or that seem most important to you. Discuss your choices with the class. As you do so, consider how in this poem Poe creates a language of his own to make a timeless topic unique. How attuned should health care providers be to various styles of communication and points of view? How can this awareness contribute positively to patient care?

Berenice

March 1835, *Southern Literary Messenger*

Readers complained to the editor of the Southern Literary Messenger *after reading the gruesome "Berenice" in the March 1835 issue. Slyly apologizing for his "sin," Poe wrote the editor, Thomas Willis White, promising not to shock readers with such violence and gore again; however, he was also not shy about claiming that his goal was to be read and suggesting that the use of sensationalism would help him accomplish just that (Poe 1835).[1] Although he claimed to be contrite, in "Berenice" Poe actually touches upon several motifs that he would revisit throughout his career, including the apparent death of a beautiful woman and premature burial (Weekes 2002, 149).*

In this anthology, "Berenice" and "The Fall of the House of Usher" represent a selection of works based on the death of a beautiful woman, including the tales "Ligeia," "Morella," and "Eleonora," and the poems "Annabelle Lee" and "The Raven." The death-of-a-beautiful-woman motif invites questions about gender issues as they pertain to women's health. Of these stories, "Berenice" distinguishes itself because of the dentistry element and the reference to the family physician, whose medical tools are used to commit a hideous crime. Although the doctor has no speaking role, the story nonetheless urges us to ask questions about doctors' and scientists' attitudes toward and treatment of women and their bodies during the Victorian era, when the Cult of True Womanhood reigned and "hysteria" was a catchall diagnosis used to explain many female illnesses, from headaches to catatonic states (Smith-Rosenberg 1972).[2]

Suffering from a disease described by the tale's narrator as "a species of epilepsy not infrequently terminating in trance *itself," Berenice becomes the focus of her fiancé's monomaniacal obsession, and her gradually wasting body underscores her helplessness in the face of illness. The narrator's preoccupation with her physical deterioration—and his fixation on the one part of her that seems to remain healthy, her teeth—introduces important questions about gender and illness, particularly the female body and identity. At the same time, his monomaniacal obsession introduces themes of mental illness and pathological behavior that would be present in many of Poe's later tales. Thus, "Berenice" is arguably the first in a series of tales in which Poe indicts the obsessive male gaze that dominated both social and medical settings.*

Finally, since it was likely inspired in part by a February 1883 Baltimore Saturday Visiter *account of grave robbers searching for human teeth to give dentists (Campbell 1933), the tale might not only stimulate discussion of human anxieties about mortality and life after death but also inspire a history lesson about nineteenth-century grave-robbing practices. Other potential influences, namely the dental-themed tales "An Event in the Life of a Dentist," published in the* New York Mirror *(April 1833), "The Death's Head," published in the popular horror anthology* Phantasmagoriana *(Forclaz 1968), and Dr. Benjamin Rush's reports of teeth extractions (Sloane 1966), make relevant conversations about the sociohistorical evolution of dental medicine in America as well as its representations in both fiction and nonfiction across the centuries and around the globe.*

•

Dicebant mihi sodales, si sepulchrum amicæ visitarem,
curas meas aliquantulum fore levatas.[3]—EBN ZAIAT.

MISERY is manifold. The wretchedness of earth is multiform. Overreaching the wide horizon as the rainbow, its hues are as various as the hues of that arch,—as distinct too, yet as intimately blended. Overreaching the wide horizon as the rainbow! How is it that from beauty I have derived a type of unloveliness?—from the covenant of peace, a simile of sorrow? But as, in ethics, evil is a consequence of good, so, in fact, out of joy is sorrow born. Either the memory of past bliss is the anguish of to-day, or the agonies which *are,* have their origin in the ecstasies which *might have been.*

My baptismal name is Egæus; that of my family I will not mention. Yet there are no towers in the land more time-honored than my gloomy, gray, hereditary halls. Our line has been called a race of visionaries; and in many striking particulars—in the character of the family mansion—in the frescos of the chief saloon—in the tapestries of the dormitories—in the chiselling of some buttresses in the armory—but more especially in the gallery of antique paintings—in the fashion of the library chamber—and, lastly, in the very peculiar nature of the library's contents, there is more than sufficient evidence to warrant the belief.

The recollections of my earliest years are connected with that chamber, and with its volumes—of which latter I will say no more. Here died my mother. Herein was I born. But it is mere idleness to say that I had not lived before—that the soul has no previous existence. You deny it?—let us not argue the matter.

Convinced myself, I seek not to convince. There is, however, a remembrance of aërial forms—of spiritual and meaning eyes—of sounds, musical yet sad—a remembrance which will not be excluded; a memory like a shadow, vague, variable, indefinite, unsteady; and like a shadow, too, in the impossibility of my getting rid of it while the sunlight of my reason shall exist.

In that chamber was I born. Thus awaking from the long night of what seemed, but was not, nonentity, at once into the very regions of fairy-land—into a palace of imagination—into the wild dominions of monastic thought and erudition—it is not singular that I gazed around me with a startled and ardent eye—that I loitered away my boyhood in books, and dissipated my youth in reverie; but it *is* singular, that as years rolled away, and the noon of manhood found me still in the mansion of my fathers—it *is* wonderful what a stagnation there fell upon the springs of my life—wonderful how total an inversion took place in the character of my commonest thought. The realities of the world affected me as visions, and as visions only, while the wild ideas of the land of dreams became, in turn,—not the material of my every-day existence—but in very deed that existence utterly and solely in itself.

*   *   *   *   *   *   *

Berenice and I were cousins, and we grew up together in my paternal halls. Yet differently we grew—I ill of health, and buried in gloom—she agile, graceful, and overflowing with energy; hers the ramble on the hill-side—mine, the studies of the cloister—I living within my own heart, and addicted body and soul to the most intense and painful meditation—she roaming carelessly through life with no thought of the shadows in her path, or the silent flight of the raven-winged hours. Berenice!—I call upon her name—Berenice!—and from the gray ruins of memory a thousand tumultuous recollections are startled at the sound! Ah! vividly is her image before me now, as in the early days of her light-heartedness and joy! Oh! gorgeous yet fantastic beauty! Oh! sylph amid the shrubberies of Arnheim!—Oh! Naiad among its fountains!—and then—then all is mystery and terror, and a tale which should not be told. Disease—a fatal disease—fell like the simoom upon her frame, and, even while I gazed upon her, the spirit of change swept over her, pervading her mind, her habits, and her character, and, in a manner the most subtle and terrible, disturbing even the identity of her person! Alas! the destroyer came and went, and the victim—where was she? I knew her not—or knew her no longer as Berenice.

Among the numerous train of maladies superinduced by that fatal and primary one which effected a revolution of so horrible a kind in the moral and physical being of my cousin, may be mentioned as the most distressing and obstinate in its nature, a species of epilepsy not unfrequently terminating in *trance* itself—trance very nearly resembling positive dissolution, and from which her manner of recovery was, in most instances, startlingly abrupt. In the mean time my own disease—for I have been told that I should call it by no other appellation—my own disease, then, grew rapidly upon me, and assumed finally a monomaniac character of a novel and extraordinary form—hourly and momently gaining vigor—and at length obtaining over me the most incomprehensible ascendancy. This monomania, if I must so term it, consisted in a morbid irritability of those properties of the mind in metaphysical science termed the *attentive*. It is more than probable that I am not understood; but I fear, indeed, that it is in no manner possible to convey to the mind of the merely general reader, an adequate idea of that nervous *intensity of interest* with which, in my case, the powers of meditation (not to speak technically) busied and buried themselves, in the contemplation of even the most ordinary objects of the universe.

To muse for long unwearied hours with my attention riveted to some frivolous device on the margin, or in the typography of a book; to become absorbed, for the better part of a summer's day, in a quaint shadow falling aslant upon the tapestry or upon the door; to lose myself for an entire night in watching the steady flame of a lamp, or the embers of a fire; to dream away whole days over the perfume of a flower; to repeat monotonously some common word, until the sound, by dint of frequent repetition, ceased to convey any idea whatever to the mind; to lose all sense of motion or physical existence, by means of absolute bodily quiescence long and obstinately persevered in;—such were a few of the most common and least pernicious vagaries induced by a condition of the mental faculties, not, indeed, altogether unparalleled, but certainly bidding defiance to anything like analysis or explanation.

Yet let me not be misapprehended.—The undue, earnest, and morbid attention thus excited by objects in their own nature, frivolous, must not be confounded in character with that ruminating propensity common to all mankind, and more especially indulged in by persons of ardent imagination. It was not even, as might be at first supposed, an extreme condition, or exaggeration of such propensity, but primarily and essentially distinct and different. In the one instance, the dreamer, or enthusiast, being interested by an object usually *not* frivolous, imperceptibly loses sight of this object in a wilderness of deductions and suggestions issuing therefrom, until, at the conclusion of a day dream *often replete with luxury,* he finds the *incitamentum,* or first cause of his musings, entirely vanished and forgotten. In my case, the primary object was *invariably frivolous,* although assuming, through the medium of my distempered vision, a refracted and unreal importance. Few deductions, if any, were made; and those few pertinaciously returning in upon the original object as a centre. The meditations were *never* pleasurable; and at the termination of the reverie the first cause, so far from being out of sight, had attained that supernaturally exaggerated interest which was the prevailing feature of the disease. In a word, the powers of mind more particularly exercised were, with me, as I have said before, the *attentive,* and are, with the day-dreamer, the *speculative.*

My books, at this epoch, if they did not actually serve to irritate the disorder, partook, it will be perceived, largely, in their imaginative and inconsequential nature, of the characteristic qualities of the disorder itself. I well remember, among others, the treatise of the noble Italian, Cœlius Secundus Curio, *"de Amplitudine Beati Regni Dei;"* St. Austin's great work, "City of God;" and Tertullian *"de Carne Christi,"* in which the paradoxical sentence, *"Mortuus est Dei filius; credibile est quia ineptum est: et sepultus resurrexit; certum est quia impossibile est"* occupied my undivided time, for many weeks of laborious and fruitless investigation.[4]

Thus it will appear that, shaken from its balance only by trivial things, my reason bore resemblance to that ocean-crag spoken of by Ptolemy Hephestion, which steadily resisting the attacks of human violence, and the fiercer fury of the waters and the winds, trembled only to the touch of the flower called Asphodel. And although, to a careless thinker, it might appear a matter beyond doubt, that the alteration produced by her unhappy malady, in the *moral* condition of Berenice, would afford me many objects for the exercise of that intense and abnormal meditation whose nature I have been at some trouble in explaining, yet such was not in any degree the case. In the lucid intervals of my infirmity, her calamity, indeed, gave me pain, and, taking deeply to heart that total wreck of her fair and gentle life, I did not fail to ponder, frequently and bitterly, upon the wonder-working means by which so strange a revolution had been so suddenly brought to pass. But these reflections partook not of the idiosyncrasy of my disease, and were such as would have occurred, under similar circumstances, to the ordinary mass of mankind. True to its own character, my disorder revelled in the less important but more startling changes wrought in the *physical* frame of Berenice—in the singular and most appalling distortion of her personal identity.

During the brightest days of her unparalleled beauty, most surely I had never loved her. In the strange anomaly of my existence, feelings with me, *had never been* of the heart, and my passions *always were* of the mind. Through the gray of the early morning—among the trellised shadows of the forest at noonday—and in the silence of my library at night—she had flitted by my eyes, and I had seen her—not as the living and breathing Berenice, but as the Berenice of a dream; not as a being of the earth, earthy, but as the abstraction of such a being—not as a thing to admire, but to analyze—not as an object of love, but as the theme of the most abstruse although desultory speculation. And *now*—now I shuddered in her presence, and grew pale at her approach; yet bitterly lamenting her fallen and desolate condition, I called to mind that she had loved me long, and, in an evil moment, I spoke to her of marriage.

And at length the period of our nuptials was approaching, when, upon an afternoon in the winter of the year,—one of those unseasonably warm, calm, and misty days which are the nurse of the beautiful Halcyon,* I sat (and sat, as I thought, alone) in the inner apartment of the library. But, uplifting my eyes, I saw that Berenice stood before me.

Was it my own excited imagination—or the misty influence of the atmosphere—or the uncertain twilight of the chamber—or the gray draperies which fell around her figure—that caused in it so vacillating and indistinct an outline? I could not tell. She spoke no word, and I—not for worlds could I have uttered a syllable. An icy chill ran through my frame; a sense of insufferable anxiety oppressed me; a consuming curiosity pervaded my soul; and sinking back upon the chair, I remained for some time breathless and motionless, with my eyes riveted upon her person. Alas! its emaciation was excessive, and not one vestige of the former being lurked in any single line of the contour. My burning glances at length fell upon the face.

The forehead was high, and very pale, and singularly placid; and the once jetty hair fell partially over it, and overshadowed the hollow temples with innumerable ringlets, now of a vivid yellow, and jarring discordantly, in their fantastic character, with the reigning melancholy of the countenance. The eyes were lifeless, and lustreless, and seemingly pupil-less, and I shrank involuntarily from their glassy stare to the contemplation of the thin and shrunken lips. They parted; and in a smile of peculiar meaning, *the teeth* of the changed Berenice disclosed themselves slowly to my view. Would to God that I had never beheld them, or that, having done so, I had died!

*   *   *   *   *   *

*For as Jove, during the winter season, gives twice seven days of warmth, men have called this element and temperate time the nurse of the beautiful Halcyon.—*Simonides.*

The shutting of a door disturbed me, and, looking up, I found that my cousin had departed from the chamber. But from the disordered chamber of my brain, had not, alas! departed, and would not be driven away, the white and ghastly *spectrum* of the teeth. Not a speck on their surface—not a shade on their enamel—not an indenture in their edges—but what that period of her smile had sufficed to brand in upon my memory. I saw them *now* even more unequivocally than I beheld them *then*. The teeth!—the teeth!—they were here, and there, and every where, and visibly and palpably before me; long, narrow, and excessively white, with the pale lips writhing about them, as in the very moment of their first terrible development. Then came the full fury of my *monomania,* and I struggled in vain against its strange and irresistible influence. In the multiplied objects of the external world I had no thoughts but for the teeth. For these I longed with a phrenzied desire. All other matters and all different interests became absorbed in their single contemplation. They—they alone were present to the mental eye, and they, in their sole individuality, became the essence of my mental life. I held them in every light. I turned them in every attitude. I surveyed their characteristics. I dwelt upon their peculiarities. I pondered upon their conformation. I mused upon the alteration in their nature. I shuddered as I assigned to them in imagination a sensitive and sentient power, and even when unassisted by the lips, a capability of moral expression. Of Mad'selle Sallé it has been well said, *"que tous ses pas étaient des sentiments,"* and of Berenice I more seriously believed *que toutes ses dents étaient des idées. Des idées!*—ah here was the idiotic thought that destroyed me! *Des idées!*—ah *therefore* it was that I coveted them so madly![5] I felt that their possession could alone ever restore me to peace, in giving me back to reason.

And the evening closed in upon me thus—and then the darkness came, and tarried, and went—and the day again dawned—and the mists of a second night were now gathering around—and still I sat motionless in that solitary room; and still I sat buried in meditation, and still the *phantasma* of the teeth maintained its terrible ascendancy as, with the most vivid and hideous distinctness, it floated about amid the changing lights and shadows of the chamber. At length there broke in upon my dreams a cry as of horror and dismay; and thereunto, after a pause, succeeded the sound of troubled voices, intermingled with many low moanings of sorrow, or of pain. I arose from my seat and, throwing open one of the doors of the library, saw standing out in the ante-chamber a servant maiden, all in tears, who told me that Berenice was—no more. She had been seized with epilepsy in the early morning, and now, at the closing in of the night, the grave was ready for its tenant, and all the preparations for the burial were completed.

*   *   *   *   *   *   *

I found myself sitting in the library, and again sitting there alone. It seemed to me that I had newly awakened from a confused and exciting dream. I knew that it was now midnight, and I was well aware that since the setting of the sun Berenice had been interred. But of that dreary period which intervened I had no positive, at least no definite, comprehension. Yet its memory was replete with horror—horror more horrible from being vague, and terror more terrible from ambiguity. It was a fearful page in the record of my existence, written all over with dim, and hideous, and unintelligible recollections. I strived to decipher them, but in vain; while ever and anon, like the spirit of a departed sound, the shrill and piercing shriek of a female voice seemed to be ringing in my ears. I had done a deed—what was it? I asked myself the question aloud, and the whispering echoes of the chamber answered me—*"what was it?"*

On the table beside me burned a lamp, and near it lay a little box. It was of no remarkable character, and I had seen it frequently before, for it was the property of the family physician; but how came it *there,* upon my table, and why did I shudder in regarding it? These things were in no manner to be accounted for, and my eyes at length dropped to the open pages of a book, and to a sentence underscored therein. The words were the singular but simple ones of the poet Ebn Zaiat, *"Dicebant mihi sodales si sepulchrum amicæ visitarem, curas meas aliquantulum fore levatas."* Why, then, as I perused them, did the hairs of my head erect themselves on end, and the blood of my body become congealed within my veins?

There came a light tap at the library door, and pale as the tenant of a tomb, a menial entered upon tiptoe. His looks were wild with terror, and he spoke to me in a voice tremulous, husky, and very low. What said he?—some broken sentences I heard. He told of a wild cry disturbing the silence of the night—of the gathering together of the household—of a search in the direction of the sound; and then his tones grew thrillingly distinct as he whispered me of a violated grave—of a disfigured body enshrouded, yet still breathing—still pal-pitating—still *alive!*

He pointed to garments;—they were muddy and clotted with gore. I spoke not, and he took me gently by the hand; it was indented with the impress of human nails. He directed my attention to some object against the wall;—I looked at it for some minutes;—it was a spade. With a shriek I bounded to the table, and grasped the box that lay upon it. But I could not force it open; and in my tremor it slipped from my hands, and fell heavily, and burst into pieces;

and from it, with a rattling sound, there rolled out some instruments of dental surgery, intermingled with thirty-two small, white and ivory-looking substances that were scattered to and fro about the floor.

DISCUSSION QUESTIONS

1. Describe Berenice and Egæus. Are they compatible? What does Egæus value about Berenice, and how does that valuation change after she falls ill?

2. Analyze the passage in which Egæus describes his growing obsession with Berenice's teeth. To what do you attribute this? What do the teeth represent to him? Research the terms "idée fixe" and "monomania," and explain how they relate to these questions. In what ways does this tale speak to the dangers of applying a cold, objective gaze, like the one introduced by the new clinical medicine imported to America in the middle of the nineteenth century?

3. Compare Egæus's obsession with Berenice's teeth to Aylmer's obsession with Georgiana's birthmark in Hawthorne's "The Birthmark." Because nineteenth-century physicians who practiced the new clinical medicine valued physical diagnosis but, out of propriety, often avoided viewing the female body—even when performing obstetrical and gynecological procedures (Browner 2005)—could the female body, in particular, have more value in death than in life? In what ways might this inform our readings of the two tales?

4. With the rise of clinical medicine, when diagnosing or healing patients many physicians relied less often on patient narratives and more on physical examination and clinical observation. Is it significant, then, that Berenice does not have a voice in telling her own story of illness? What techniques might a health care provider use today to elicit the particulars of a patient's story? What advice do narrative health care scholars offer about developing listening and interpreting skills?

5. Although the family physician does not appear in the tale, his tools are used in Egæus's assault on Berenice's body. What significance might we assign to this detail? Is the physician, or the medical community he represents, in some way culpable for the crimes committed against Berenice? Why or why not?

6. Critics and biographers often cite the loss of Poe's mother, and the subsequent loss of other female loves, as igniting Poe's fascination with the death of a beautiful woman, which he called "the most poetical topic in the world" (as quoted in Weekes 2002, 148). A number of scholars argue that Poe is not complicit with his monomaniacal narrators but motivated by "a desire to empower

the hushed woman" (Elbert 2004a, 21). Was Poe, then, subverting notions of the feminine ideal that suggested "true" women should be pious, pure, submissive, and domestic (Welter 1966)? Locate evidence in the tale to support your position.

7. Give Berenice a voice. How would she relate the details of her illness and suffering? Write a one- to two-page monologue from her perspective. Or complete this assignment for the physician, who also has no voice in the tale.

The Fall of the House of Usher
September 1839, *Burton's Gentleman's Magazine*

Nineteenth-century phrenologists prohibited incest, describing it as an abuse "of the amative propensity" (Spurzheim 1825, 170); along with physiologists, they perceived incest as a public health issue, believing it caused the "physical and mental degeneration of offspring" (Connolly 2014, 123). In contrast, some physiologists believed inbreeding could actually improve the gene pool, resulting in superior progeny. Debating incest on reproductive grounds, these medico-scientific men added new dimension to the incest war that theologians and others waged on moral grounds.[1] America's anxiety over incest is reflected in sentimental and gothic literature of the time, in which writers frequently employed it as a motif. Poe was no exception.

Poe, who both drew upon and satirized these genres, understood the appeal of the motif and likely chose it knowing it would be highly marketable. In "The Fall of the House of Usher," however, he also draws our attention to the position of the phrenologists and physiologists: the Usher family is a cautionary tale of the ill health—and ultimately destruction—that befalls incestuous nineteenth-century families and, by extension, the health of the outside community, whose members grow "infected" by association. Is it any wonder, then, that the family physician is bewildered in the face of such an illness? He offers little to no help in diagnosing or healing Madeleine or Roderick (Long 1989; Hardy 1998). Is he helpless to cure a disease that has resulted from generations of reproductive incest?[2]

Of course, brief as his appearance is, the doctor character is surrounded by other questions of interest to the health humanities. Is he helpless or plotting in his failure to diagnose and treat Madeleine, having designs on postmortem examination? Poe wrote the tale during an era when the use of cadavers in medical education was highly controversial. Published in 1839, the story emerged at the center of this controversy: between 1830 and 1850 five states passed (and later repealed because of public outcry) acts giving the medical community legal access to cadavers (Browner 2005). The public angst over these acts reflected a core, widespread fear that the medical community had begun to value individuals' dead bodies more than their living ones. Would they allow patients to die so they could cut them open and study their insides?

Some physicians, in fact, accused others of these very practices. When physicians described as "therapeutic skeptics" argued that nature should be allowed to take its course in many cases of illness, doctors who believed their job was to attempt some type of treatment in all cases accused these skeptics of being more interested in performing postmortem exams than in curing patients (Browner 2005). Poe, immersed in the world of mass media and read medical literature, was aware of these controversies. Thus, while this tale provides only a momentary glimpse of the doctor, it invites us to speculate about the doctor's various motivations—and to analyze the attempts other characters make to "play doctor" in his absence.

•

"Son cœur est un luth suspendu; Sitôt qu'on le touche il résonne."[3]
—De Béranger.

DURING the whole of a dull, dark, and soundless day in the autumn of the year, when the clouds hung oppressively low in the heavens, I had been passing alone, on horseback, through a singularly dreary tract of country; and at length found myself, as the shades of the evening drew on, within view of the melancholy House of Usher. I know not how it was —but, with the first glimpse of the building, a sense of insufferable gloom pervaded my spirit. I say insufferable; for the feeling was unrelieved by any of that half-pleasurable, because poetic, sentiment, with which the mind usually receives even the sternest natural images of the desolate or terrible. I looked upon the scene before me—upon the mere house, and the simple landscape features of the domain—upon the bleak walls—upon the vacant eye-like windows—upon a few rank sedges— and upon a few white trunks of decayed trees—with an utter depression of soul which I can compare to no earthly sensation more properly than to the after-dream of the reveller upon opium—the bitter lapse into everyday life—the hideous dropping off of the veil. There was an iciness, a sinking, a sickening of the heart—an unredeemed dreariness of thought which no goading of the imagination could torture into aught of the sublime. What was it—I paused to think—what was it that so unnerved me in the contemplation of the House of Usher? It was a mystery all insoluble; nor could I grapple with the shadowy fancies that crowded upon me as I pondered. I was forced to fall back upon the unsatisfactory conclusion, that while, beyond doubt, there *are* combinations of very simple natural objects which have the power of thus affecting us, still the analysis of this power lies among considerations beyond our depth. It was possible, I reflected, that a mere different arrangement of the particulars of the scene, of the details of the picture, would be sufficient to modify, or perhaps to annihilate its capacity for sorrowful impression; and, acting upon this idea, I reined my horse to the precipitous brink of a black and lurid tarn that lay in unruffled lustre by the dwelling, and gazed down—but with a shudder even more thrilling than before—upon the remodelled and inverted images of the gray sedge, and the ghastly tree-stems, and the vacant and eye-like windows.

Nevertheless, in this mansion of gloom I now proposed to myself a sojourn of some weeks. Its proprietor, Roderick Usher, had been one of my boon companions in boyhood; but many years had elapsed since our last meeting. A letter, however, had lately reached me in a distant part of the country—a letter from him—which, in its wildly importunate nature, had admitted of no other than a personal reply. The MS. gave evidence of nervous agitation. The writer spoke of acute bodily illness—of a mental disorder which oppressed him—and of an earnest desire to see me, as his best and indeed his only personal friend, with a view of attempting, by the cheerfulness of my society, some alleviation of his malady. It was the manner in which all this, and much more, was said—it was the apparent *heart* that went with his request—which allowed me no room for hesitation; and I accordingly obeyed forthwith what I still considered a very singular summons.

Although, as boys, we had been even intimate associates, yet I really knew little of my friend. His reserve had been always excessive and habitual. I was aware, however, that his very ancient family had been noted, time out of mind, for a peculiar sensibility of temperament, displaying itself, through long ages, in many works of exalted art, and manifested, of late, in repeated deeds of munificent yet unobtrusive charity, as well as in a passionate devotion to the intricacies, perhaps even more than to the orthodox and easily recognisable beauties, of musical science. I had learned, too, the very remarkable fact, that the stem of the Usher race, all time-honored as it was, had put forth, at no period, any enduring branch; in other words, that the entire family lay in the direct line of descent, and had always, with very trifling and very temporary variation, so lain. It was this deficiency, I considered, while running over in thought the perfect keeping of the character of the premises with the accredited character of the people, and while speculating upon the possible influence which the one, in the long lapse of centuries, might have exercised upon the other—it was this deficiency, perhaps, of collateral issue, and the consequent undeviating transmission, from sire to son, of the patrimony with the name, which had, at length, so identified the two as to merge the original title of the estate in the quaint and equivocal appellation of the "House of Usher"—an appellation which seemed to include, in the minds of the peasantry who used it, both the family and the family mansion.

I have said that the sole effect of my somewhat childish experiment—that of looking down within the tarn—had been to deepen the first singular impression. There can be no doubt that the consciousness of the rapid increase of my superstition—for why should I not so term it?—served mainly to accelerate the increase itself. Such, I have long known, is the paradoxical law of all sentiments having terror as a basis. And it might have been for this reason only, that, when I again uplifted my eyes to the house itself, from its image in the pool, there grew in my mind a strange fancy—a fancy so ridiculous, indeed, that I but mention it to show the vivid force of the sensations which oppressed me. I had so worked upon my imagination as really to believe that about the whole mansion and domain there hung an atmosphere peculiar to themselves and their immediate vicinity—an atmosphere which had no affinity with the air of heaven, but which had reeked up from the decayed trees, and the gray wall, and the silent tarn—a pestilent and mystic vapor, dull, sluggish, faintly discernible, and leaden-hued.

Shaking off from my spirit what *must* have been a dream, I scanned more narrowly the real aspect of the building. Its principal feature seemed to be that of an excessive antiquity. The discoloration of ages had been great. Minute fungi overspread the whole exterior, hanging in a fine tangled web-work from the eaves. Yet all this was apart from any extraordinary dilapidation. No portion of the masonry had fallen; and there appeared to be a wild inconsistency between its still perfect adaptation of parts, and the crumbling condition of the individual stones. In this there was much that reminded me of the specious totality of old wood-work which has rotted for long years in some neglected vault, with no disturbance from the breath of the external air. Beyond this indication of extensive decay, however, the fabric gave little token of instability. Perhaps the eye of a scrutinizing observer might have discovered a barely perceptible fissure, which, extending from the roof of the building in front, made its way down the wall in a zigzag direction, until it became lost in the sullen waters of the tarn.

Noticing these things, I rode over a short causeway to the house. A servant in waiting took my horse, and I entered the Gothic archway of the hall. A valet, of stealthy step, thence conducted me, in silence, through many dark and intricate passages in my progress to the *studio* of his master. Much that I encountered on the way contributed, I know not how, to heighten the vague sentiments of which I have already spoken. While the objects around me—while the carvings of the ceilings, the sombre tapestries of the walls, the ebon blackness of the floors, and the phantasmagoric armorial trophies which rattled as I strode, were but matters to which, or to such as which, I had been accustomed from my infancy—while I hesitated not to acknowledge how familiar was all this—I still wondered to find how unfamiliar were the fancies which ordinary images were stirring up. On one of the staircases, I met the physician of the family. His countenance, I thought, wore a mingled expression of low cunning and perplexity. He accosted me with trepidation and passed on. The valet now threw open a door and ushered me into the presence of his master.

The room in which I found myself was very large and lofty. The windows were long, narrow, and pointed, and at so vast a distance from the black oaken floor as to be altogether inaccessible from within. Feeble gleams of encrimsoned light made their way through the trellissed panes, and served to render sufficiently distinct the more prominent objects around; the eye, however, struggled in vain to reach the remoter angles of the chamber, or the recesses of the vaulted and fretted ceiling. Dark draperies hung upon the walls. The general furniture was profuse, comfortless, antique, and tattered. Many books and musical instruments lay scattered about, but failed to give any vitality to the scene. I felt that I breathed an atmosphere of sorrow. An air of stern, deep, and irredeemable gloom hung over and pervaded all.

Upon my entrance, Usher arose from a sofa on which he had been lying at full length, and greeted me with a vivacious warmth which had much in it, I at first thought, of an overdone cordiality—of the constrained effort of the *ennuyé* man of the world. A glance, however, at his countenance convinced me of his perfect sincerity. We sat down; and for some moments, while he spoke not, I gazed upon him with a feeling half of pity, half of awe. Surely, man had never before so terribly altered, in so brief a period, as had Roderick Usher! It was with difficulty that I could bring myself to admit the identity of the wan being before me with the companion of my early boyhood. Yet the character of his face had been at all times remarkable. A cadaverousness of complexion; an eye large, liquid, and luminous beyond comparison; lips somewhat thin and very pallid but of a surpassingly beautiful curve; a nose of a delicate Hebrew model, but with a breadth of nostril unusual in similar formations; a finely moulded chin, speaking, in its want of prominence, of a want of moral energy; hair of a more than web-like softness and tenuity; these features, with an inordinate expansion above the regions of the temple, made up altogether a countenance not easily to be forgotten. And now in the mere exaggeration of the prevailing character of these features, and of the expression they were wont to convey, lay so much of change that I doubted to whom I spoke. The now ghastly pallor of the skin, and the now miraculous lustre of the eye, above all things startled and even awed me. The silken hair, too, had been suffered to grow all unheeded, and as, in its wild gossamer texture, it floated rather than fell about the face, I could not, even with effort, connect its Arabesque expression with any idea of simple humanity.

In the manner of my friend I was at once struck with an incoherence—an inconsistency; and I soon found this to arise from a series of feeble and futile struggles to overcome an habitual trepidancy—an excessive nervous agitation. For something of this nature I had indeed been prepared, no less by his letter, than by reminiscences of certain boyish traits, and by conclusions deduced from his peculiar physical conformation and temperament. His action was alternately vivacious and sullen. His voice varied rapidly from a tremulous indecision (when the animal spirits seemed utterly in abeyance) to that species of energetic concision—that abrupt, weighty, unhurried, and hollow-sounding enunciation—that leaden, self-balanced, and perfectly modulated guttural utterance, which may be observed in the lost drunkard, or the irreclaimable eater of opium, during the periods of his most intense excitement.

It was thus that he spoke of the object of my visit, of his earnest desire to see me, and of the solace he expected me to afford him. He entered, at some length, into what he conceived to be the nature of his malady. It was, he said, a constitutional and a family evil, and one for which he despaired to find a remedy—a mere nervous affection, he immediately added, which would undoubtedly soon pass off. It displayed itself in a host of unnatural sensations. Some of these, as he detailed them, interested and bewildered me; although, perhaps, the terms and the general manner of their narration had their weight. He suffered much from a morbid acuteness of the senses; the most insipid food was alone endurable; he could wear only garments of certain texture; the odors of all flowers were oppressive; his eyes were tortured by even a faint light; and there were but peculiar sounds, and these from stringed instruments, which did not inspire him with horror.

To an anomalous species of terror I found him a bounden slave. "I shall perish," said he, "I *must* perish in this deplorable folly. Thus, thus, and not otherwise, shall I be lost. I dread the events of the future, not in themselves, but in their results. I shudder at the thought of any, even the most trivial, incident, which may operate upon this intolerable agitation of soul. I have, indeed, no abhorrence of danger, except in its absolute effect—in terror. In this unnerved—in this pitiable condition—I feel that the period will sooner or later arrive when I must abandon life and reason together, in some struggle with the grim phantasm, FEAR."

I learned, moreover, at intervals, and through broken and equivocal hints, another singular feature of his mental condition. He was enchained by certain superstitious impressions in regard to the dwelling which he tenanted, and whence, for many years, he had never ventured forth—in regard to an influence whose supposititious force was conveyed in terms too shadowy here to be re-stated—an influence which some peculiarities in the mere form and substance of his family mansion had, by dint of long sufferance, he said, obtained over his spirit—an effect which the *physique* of the gray walls and turrets, and of the dim tarn into which they all looked down, had, at length, brought about upon the *morale* of his existence.

He admitted, however, although with hesitation, that much of the peculiar gloom which thus afflicted him could be traced to a more natural and far more palpable origin—to the severe and long-continued illness—indeed to the evidently approaching dissolution—of a tenderly beloved sister—his sole companion for long years—his last and only relative on earth. "Her decease," he said, with a bitterness which I can never forget, "would leave him (him the hopeless and the frail) the last of the ancient race of the Ushers." While he spoke, the lady Madeline (for so was she called) passed through a remote portion of the apartment, and, without having noticed my presence, disappeared. I regarded her with an utter astonishment not unmingled with dread— and yet I found it impossible to account for such feelings. A sensation of stupor oppressed me, as my eyes followed her retreating steps. When a door, at length, closed upon her, my glance sought instinctively and eagerly the countenance of the brother—but he had buried his face in his hands, and I could only perceive that a far more than ordinary wanness had overspread the emaciated fingers through which trickled many passionate tears.

The disease of the lady Madeline had long baffled the skill of her physicians. A settled apathy, a gradual wasting away of the person, and frequent although transient affections of a partially cataleptical character, were the unusual diagnosis. Hitherto she had steadily borne up against the pressure of her malady, and had not betaken herself finally to bed; but, on the closing in of the evening of my arrival at the house, she succumbed (as her brother told me at night with inexpressible agitation) to the prostrating power of the destroyer; and I learned that the glimpse I had obtained of her person would thus probably be the last I should obtain—that the lady, at least while living, would be seen by me no more.

For several days ensuing, her name was unmentioned by either Usher or myself: and during this period I was busied in earnest endeavors to alleviate the melancholy of my friend. We painted and read together; or I listened, as if in a dream, to the wild improvisations of his speaking guitar. And thus, as a closer and still closer intimacy admitted me more unreservedly into the recesses of his spirit, the more bitterly did I perceive the futility of all attempt at cheering a mind from which darkness, as if an inherent positive quality, poured forth upon all objects of the moral and physical universe, in one unceasing radiation of gloom.

I shall ever bear about me a memory of the many solemn hours I thus spent alone with the master of the House of Usher. Yet I should fail in any attempt to convey an idea of the exact character of the studies, or of the occupations, in which he involved me, or led me the way. An excited and highly distempered ideality threw a sulphureous lustre over all. His long improvised dirges will ring forever in my ears. Among other things, I hold painfully in mind a certain singular perversion and amplification of the wild air of the last waltz of Von Weber. From the paintings over which his elaborate fancy brooded, and which grew, touch by touch, into vaguenesses at which I shuddered the more thrillingly, because I shuddered knowing not why;—from these paintings (vivid as their images now are before me) I would in vain endeavor to educe more than a small portion which should lie within the compass of merely written words. By the utter simplicity, by the nakedness of his designs, he arrested and overawed attention. If ever mortal painted an idea, that mortal was Roderick Usher. For me at least—in the circumstances then surrounding me—there arose out of the pure abstractions which the hypochondriac contrived to throw upon his canvas, an intensity of intolerable awe, no shadow of which felt I ever yet in the contemplation of the certainly glowing yet too concrete reveries of Fuseli.

One of the phantasmagoric conceptions of my friend, partaking not so rigidly of the spirit of abstraction, may be shadowed forth, although feebly, in words. A small picture presented the interior of an immensely long and rectangular vault or tunnel, with low walls, smooth, white, and without interruption or device. Certain accessory points of the design served well to convey the idea that this excavation lay at an exceeding depth below the surface of the earth. No outlet was observed in any portion of its vast extent, and no torch, or other artificial source of light was discernible; yet a flood of intense rays rolled throughout, and bathed the whole in a ghastly and inappropriate splendor.

I have just spoken of that morbid condition of the auditory nerve which rendered all music intolerable to the sufferer, with the exception of certain effects of stringed instruments. It was, perhaps, the narrow limits to which he thus confined himself upon the guitar, which gave birth, in great measure, to the fantastic character of his performances. But the fervid *facility* of his *impromptus* could not be so accounted for. They must have been, and were, in the notes, as well as in the words of his wild fantasias (for he not unfrequently accompanied himself with rhymed verbal improvisations), the result of that intense mental collectedness and concentration to which I have previously alluded as observable only in particular moments of the highest artificial excitement. The words of one of these rhapsodies I have easily remembered. I was, perhaps, the more forcibly impressed with it as he gave it, because, in the under or mystic current of its meaning, I fancied that I perceived, and for the first time, a full consciousness on the part of Usher of the tottering of his lofty reason upon her throne. The verses, which were entitled "The Haunted Palace," ran very nearly, if not accurately, thus:

I.

In the greenest of our valleys,
    By good angels tenanted,
Once a fair and stately palace—
    Radiant palace—reared its head.
In the monarch Thought's dominion—
    It stood there!
Never seraph spread a pinion
    Over fabric half so fair.

II.

Banners yellow, glorious, golden,
    On its roof did float and flow;
(This—all this—was in the olden
    Time long ago)
And every gentle air that dallied,
    In that sweet day,
Along the ramparts plumed and pallid,
    A winged odour went away.

III.

Wanderers in that happy valley
    Through two luminous windows saw
Spirits moving musically
    To a lute's well-tunèd law;
Round about a throne, where sitting
    (Porphyrogene!)
In state his glory well befitting,
    The ruler of the realm was seen.

IV.

And all with pearl and ruby glowing
    Was the fair palace door,
Through which came flowing, flowing, flowing
    And sparkling evermore,
A troop of Echoes whose sweet duty
    Was but to sing,
In voices of surpassing beauty,
    The wit and wisdom of their king.

V.

But evil things, in robes of sorrow,
    Assailed the monarch's high estate;
(Ah, let us mourn, for never morrow
    Shall dawn upon him, desolate!)
And, round about his home, the glory
    That blushed and bloomed
Is but a dim-remembered story
    Of the old time entombed.

VI.

And travellers now within that valley,
    Through the red-litten windows, see
Vast forms that move fantastically
    To a discordant melody;
While, like a rapid ghastly river,
    Through the pale door,
A hideous throng rush out forever,
    And laugh—but smile no more.

I well remember that suggestions arising from this ballad, led us into a train of thought wherein there became manifest an opinion of Usher's which I mention not so much on account of its novelty (for other men* have thought thus), as on account of the pertinacity with which he maintained it. This opinion, in its general form, was that of the sentience of all vegetable things. But, in his disordered fancy, the idea had assumed a more daring character, and

---

*Watson, Dr. Percival, Spallanzani, and especially of the Bishop of Landaff—See "Chemical Essays," vol. v.

trespassed, under certain conditions, upon the kingdom of inorganization. I lack words to express the full extent, or the earnest *abandon* of his persuasion. The belief, however, was connected (as I have previously hinted) with the gray stones of the home of his forefathers. The conditions of the sentience had been here, he imagined, fulfilled in the method of collocation of these stones—in the order of their arrangement, as well as in that of the many *fungi* which overspread them, and of the decayed trees which stood around—above all, in the long undisturbed endurance of this arrangement, and in its reduplication in the still waters of the tarn. Its evidence—the evidence of the sentience—was to be seen, he said (and I here started as he spoke), in the gradual yet certain condensation of an atmosphere of their own about the waters and the walls. The result was discoverable, he added, in that silent yet importunate and ter-rible influence which for centuries had moulded the destinies of his family, and which made *him* what I now saw him—what he was. Such opinions need no comment, and I will make none.

Our books—the books which, for years, had formed no small portion of the mental existence of the invalid—were, as might be supposed, in strict keeping with this character of phantasm. We pored together over such works as the Ververt et Chartreuse of Gresset; the Belphegor of Machiavelli; the Heaven and Hell of Swedenborg; the Subterranean Voyage of Nicholas Klimm of Holberg; the Chiromancy of Robert Flud, of Jean D'Indaginé, and of De la Chambre; the Journey into the Blue Distance of Tieck; and the City of the Sun of Campanella. One favourite volume was a small octavo edition of the *Directorium Inquisitorum,* by the Dominican Eymeric de Gironne; and there were passages in Pomponius Mela, about the old African Satyrs and Ægipans, over which Usher would sit dreaming for hours. His chief delight, however, was found in the perusal of an exceedingly rare and curious book in quarto Gothic—the manual of a forgotten church—the *Vigiliæ Mortuorum secundum Chorum Ecclesiæ Maguntinæ.*

I could not help thinking of the wild ritual of this work, and of its probable influence upon the hypochondriac, when, one evening, having informed me abruptly that the lady Madeline was no more, he stated his intention of pre-serving her corpse for a fortnight (previously to its final interment), in one of the numerous vaults within the main walls of the building. The worldly reason, however, assigned for this singular proceeding, was one which I did not feel at liberty to dispute. The brother had been led to his resolution (so he told me) by consideration of the unusual character of the malady of the deceased, of certain obtrusive and eager inquiries on the part of her medical men, and of the remote and exposed situation of the burial-ground of the family. I will not deny that when I called to mind the sinister countenance of the person whom I met upon the staircase, on the day of my arrival at the house, I had no desire to oppose what I regarded as at best but a harmless, and by no means an unnatural, precaution.

At the request of Usher, I personally aided him in the arrangements for the temporary entombment. The body having been encoffined, we two alone bore it to its rest. The vault in which we placed it (and which had been so long unopened that our torches, half smothered in its oppressive atmosphere, gave us little opportunity for investigation) was small, damp, and entirely without means of admission for light; lying, at great depth, immediately beneath that portion of the building in which was my own sleeping apartment. It had been used, apparently, in remote feudal times, for the worst purposes of a donjon-keep, and, in later days, as a place of deposit for powder, or some other highly combustible substance, as a portion of its floor, and the whole interior of a long archway through which we reached it, were carefully sheathed with copper. The door, of massive iron, had been, also, similarly protected. Its immense weight caused an unusually sharp, grating sound, as it moved upon its hinges.

Having deposited our mournful burden upon tressels within this region of horror, we partially turned aside the yet unscrewed lid of the coffin, and looked upon the face of the tenant. A striking similitude between the brother and sister now first arrested my attention; and Usher, divining, perhaps, my thoughts, murmured out some few words from which I learned that the deceased and himself had been twins, and that sympathies of a scarcely intelligible nature had always existed between them. Our glances, however, rested not long upon the dead—for we could not regard her unawed. The disease which had thus entombed the lady in the maturity of youth, had left, as usual in all maladies of a strictly cataleptical character, the mockery of a faint blush upon the bosom and the face, and that suspiciously lingering smile upon the lip which is so terrible in death. We replaced and screwed down the lid, and, having secured the door of iron, made our way, with toil, into the scarcely less gloomy apartments of the upper portion of the house.

And now, some days of bitter grief having elapsed, an observable change came over the features of the mental disorder of my friend. His ordinary manner had vanished. His ordinary occupations were neglected or forgotten. He roamed from chamber to chamber with hurried, unequal, and objectless step. The pallor of his countenance had assumed, if possible, a more ghastly hue—but the luminousness of his eye had utterly gone out. The once occasional huskiness of his tone was heard no more; and a tremulous quaver, as if of extreme terror, habitually characterized his utterance. There were times, indeed, when I thought his unceasingly agitated mind was laboring with some oppressive secret, to divulge which he struggled for the necessary courage. At times, again, I was obliged to resolve all into the mere inexplicable vagaries of madness, for I beheld him gazing upon vacancy for long hours, in an attitude of the profoundest attention, as if listening to some imaginary sound. It was no wonder that his condition terrified—that it infected me. I felt creeping upon me, by slow yet certain degrees, the wild influences of his own fantastic yet impressive superstitions.

It was, especially, upon retiring to bed late in the night of the seventh or eighth day after the placing of the lady Madeline within the donjon, that I experienced the full power of such feelings. Sleep came not near my couch—while the hours waned and waned away. I struggled to reason off the nervousness which had dominion over me. I endeavoured to believe that much, if not all of what I felt, was due to the bewildering influence of the gloomy furniture of the room—of the dark and tattered draperies, which, tortured into motion by the breath of a rising tempest, swayed fitfully to and fro upon the walls, and rustled uneasily about the decorations of the bed. But my efforts were fruitless. An irrepressible tremour gradually pervaded my frame; and, at length, there sat upon my very heart an incubus of utterly causeless alarm. Shaking this off with a gasp and a struggle, I uplifted myself upon the pillows, and, peering earnestly within the intense darkness of the chamber, hearkened—I know not why, except that an instinctive spirit prompted me—to certain low and indefinite sounds which came, through the pauses of the storm, at long intervals, I knew not whence. Overpowered by an intense sentiment of horror, unaccountable yet unendurable, I threw on my clothes with haste (for I felt that I should sleep no more during the night), and endeavoured to arouse myself from the pitiable condition into which I had fallen, by pacing rapidly to and fro through the apartment.

I had taken but few turns in this manner, when a light step on an adjoining staircase arrested my attention. I presently recognized it as that of Usher. In an instant afterward he rapped, with a gentle touch, at my door, and entered, bearing a lamp. His countenance was, as usual, cadaverously wan—but, moreover, there was a species of mad hilarity in his eyes—an evidently restrained *hysteria* in his whole demeanor. His air appalled me—but anything was preferable to the solitude which I had so long endured, and I even welcomed his presence as a relief.

"And you have not seen it?" he said abruptly, after having stared about him for some moments in silence—"you have not then seen it?—but, stay! you shall." Thus speaking, and having carefully shaded his lamp, he hurried to one of the casements, and threw it freely open to the storm.

The impetuous fury of the entering gust nearly lifted us from our feet. It was, indeed, a tempestuous yet sternly beautiful night, and one wildly singular in its terror and its beauty. A whirlwind had apparently collected its force in our vicinity; for there were frequent and violent alterations in the direction of the wind; and the exceeding density of the clouds (which hung so low as to press upon the turrets of the house) did not prevent our perceiving the life-like velocity with which they flew careering from all points against each other, without passing away into the distance. I say that even their exceeding density did not prevent our perceiving this—yet we had no glimpse of the moon or stars—nor was there any flashing forth of the lightning. But the under surfaces of the huge masses of agitated vapour, as well as all terrestrial objects immediately around us, were glowing in the unnatural light of a faintly luminous and distinctly visible gaseous exhalation which hung about and enshrouded the mansion.

"You must not—you shall not behold this!" said I, shuddering, to Usher, as I led him, with a gentle violence, from the window to a seat. "These appearances, which bewilder you, are merely electrical phenomena not uncommon—or it may be that they have their ghastly origin in the rank miasma of the tarn. Let us close this casement;—the air is chilling and dangerous to your frame. Here is one of your favorite romances. I will read, and you shall listen;—and so we will pass away this terrible night together."

The antique volume which I had taken up was the "Mad Trist" of Sir Launcelot Canning; but I had called it a favourite of Usher's more in sad jest than in earnest; for, in truth, there is little in its uncouth and unimaginative prolixity which could have had interest for the lofty and spiritual ideality of my friend. It was, however, the only book immediately at hand; and I indulged a vague hope that the excitement which now agitated the hypochondriac, might find relief (for the history of mental disorder is full of similar anomalies) even in the extremeness of the folly which I should read. Could I have judged, indeed, by the wild overstrained air of vivacity with which he hearkened, or apparently hearkened, to the words of the tale, I might well have congratulated myself upon the success of my design.

I had arrived at that well-known portion of the story where Ethelred, the hero of the Trist, having sought in vain for peaceable admission into the dwelling of the hermit, proceeds to make good an entrance by force. Here, it will be remembered, the words of the narrative run thus:

"And Ethelred, who was by nature of a doughty heart, and who was now mighty withal, on account of the powerfulness of the wine which he had drunken, waited no longer to hold parley with the hermit, who, in sooth, was of an obstinate and maliceful turn, but, feeling the rain upon his shoulders, and fearing the rising of the tempest, uplifted his mace outright, and, with blows, made quickly room in the plankings of the door for his gauntleted hand; and now pulling therewith sturdily, he so cracked, and ripped, and tore all asunder, that the noise of the dry and hollow-sounding wood alarumed and reverberated throughout the forest."

At the termination of this sentence I started and, for a moment, paused; for it appeared to me (although I at once concluded that my excited fancy had deceived me)—it appeared to me that, from some very remote portion of the mansion, there came, indistinctly, to my ears, what might have been, in its exact similarity of character, the echo (but a stifled and dull one certainly) of the very cracking and ripping sound which Sir Launcelot had so particularly described. It was, beyond doubt, the coincidence alone which had arrested my attention; for, amid the rattling of the sashes of the casements, and the ordinary commingled noises of the still increasing storm, the sound, in itself, had nothing, surely, which should have interested or disturbed me. I continued the story:

"But the good champion Ethelred, now entering within the door, was sore enraged and amazed to perceive no signal of the maliceful hermit; but, in the stead thereof, a dragon of a scaly and prodigious demeanor, and of a fiery tongue, which sate in guard before a palace of gold, with a floor of silver; and upon the wall there hung a shield of shining brass with this legend enwritten—

Who entereth herein, a conqueror hath bin;
Who slayeth the dragon, the shield he shall win;

And Ethelred uplifted his mace, and struck upon the head of the dragon, which fell before him, and gave up his pesty breath, with a shriek so horrid and harsh, and withal so piercing, that Ethelred had fain to close his ears with his hands against the dreadful noise of it, the like whereof was never before heard."

Here again I paused abruptly, and now with a feeling of wild amazement—for there could be no doubt whatever that, in this instance, I did actually hear (although from what direction it proceeded I found it impossible to say) a low and apparently distant, but harsh, protracted, and most unusual screaming or grating sound—the exact counterpart of what my fancy had already conjured up for the dragon's unnatural shriek as described by the romancer.

Oppressed, as I certainly was, upon the occurrence of this second and most extraordinary coincidence, by a thousand conflicting sensations, in which wonder and extreme terror were predominant, I still retained sufficient presence of mind to avoid exciting, by any observation, the sensitive nervousness of my companion. I was by no means certain that he had noticed the sounds in question; although, assuredly, a strange alteration had, during the last few minutes, taken place in his demeanour. From a position fronting my own, he had gradually brought round his chair, so as to sit with his face to the door of the chamber; and thus I could but partially perceive his features, although I saw that his lips trembled as if he were murmuring inaudibly. His head had dropped upon his breast—yet I knew that he was not asleep, from the wide and rigid opening of the eye as I caught a glance of it in profile. The motion of his body, too, was at variance with this idea—for he rocked from side to side with a gentle yet constant and uniform sway. Having rapidly taken notice of all this, I resumed the narrative of Sir Launcelot, which thus proceeded:

"And now, the champion, having escaped from the terrible fury of the dragon, bethinking himself of the brazen shield, and of the breaking up of the enchantment which was upon it, removed the carcass from out of the way before him, and approached valorously over the silver pavement of the castle to where the shield was upon the wall; which in sooth tarried not for his full coming, but fell down at his feet upon the silver floor, with a mighty great and terrible ringing sound."

No sooner had these syllables passed my lips, than—as if a shield of brass had indeed, at the moment, fallen heavily upon a floor of silver—I became aware of a distinct, hollow, metallic, and clangorous, yet apparently muffled, reverberation. Completely unnerved, I leaped to my feet; but the measured rocking movement of Usher was undisturbed. I rushed to the chair in which he sat. His eyes were bent fixedly before him, and throughout his whole countenance there reigned a stony rigidity. But, as I placed my hand upon his shoulder, there came a strong shudder over his whole person; a sickly smile quivered about his lips; and I saw that he spoke in a low, hurried, and gibbering murmur, as if unconscious of my presence. Bending closely over him, I at length drank in the hideous import of his words.

"Now hear it?—yes, I hear it, and *have* heard it. Long—long—long—many minutes, many hours, many days, have I heard it—yet I dared not—oh, pity me, miserable wretch that I am!—I dared not—I *dared* not speak! *We have put her living in the tomb!* Said I not that my senses were acute? I *now* tell you that I heard her first feeble movements in the hollow coffin. I heard them—many, many days ago—yet I dared not—*I dared not speak!* And now—to-night—Ethelred—ha! ha!—the breaking of the hermit's door, and the death-cry of the dragon, and the clangour of the shield!—say, rather, the rending of her coffin, and the grating of the iron hinges of her prison, and her struggles within the coppered archway of the vault! Oh whither shall I fly? Will she not be here anon? Is she not hurrying to upbraid me for my haste? Have I not heard her footstep on the stair? Do I not distinguish that heavy and horrible beating of her heart? MADMAN!"—here he sprang furiously to his feet, and shrieked out his syllables, as if in the effort he were giving up his soul—"MADMAN! I TELL YOU THAT SHE NOW STANDS WITHOUT THE DOOR!"

As if in the superhuman energy of his utterance there had been found the potency of a spell—the huge antique panels to which the speaker pointed, threw slowly back, upon the instant, their ponderous and ebony jaws. It was the work of the rushing gust—but then without those doors there DID stand the lofty and enshrouded figure of the lady Madeline of Usher. There was blood upon her white robes, and the evidence of some bitter struggle upon every portion of her emaciated frame. For a moment she remained trembling and reeling to and fro upon the threshold, then, with a low moaning cry, fell heavily inward upon the person of her brother, and in her violent and now final death-agonies, bore him to the floor a corpse, and a victim to the terrors he had anticipated.

From that chamber, and from that mansion, I fled aghast. The storm was still abroad in all its wrath as I found myself crossing the old causeway. Suddenly there shot along the path a wild light, and I turned to see whence a gleam so unusual could have issued; for the vast house and its shadows were alone behind me. The radiance was that of the full, setting, and blood-red moon, which now shone vividly through that once barely-discernible fissure of which I have before spoken as extending from the roof of the building, in a zigzag direction, to the base. While I gazed, this fissure rapidly widened—there came a fierce breath of the whirlwind—the entire orb of the satellite burst at once upon my sight—my brain reeled as I saw the mighty walls rushing asunder—there was a long tumultuous shouting sound like the voice of a thousand waters—and the deep and dank tarn at my feet closed sullenly and silently over the fragments of the "HOUSE OF USHER."

DISCUSSION QUESTIONS

1. Although Madeleine is in need of medical attention, the doctor appears only very briefly in this tale. Analyze the narrator's description of the doctor and the implications of his fleeting appearance—and subsequent disappearance. In what ways do the doctor's lack of efficacy and his absence speak to cultural anxieties related to the practice of nineteenth-century medicine? Consider issues related to premature burial and therapeutic skepticism described in the introduction to this tale.

2. Roderick Usher suffers from an illness as well. Two methods of healing were used in the nineteenth century to address overly emotional patients like Roderick Usher: manly discourse and phrenology (Zimmerman 2007). Find evidence of both approaches in the story; is either successful to any degree?

3. Investigate the theme of storytelling in the tale. Does Madeleine's inability to tell her own story make her more vulnerable? Is the narrator reliable or unreliable as a storyteller? What factors might make a health care provider an unreliable narrator of a patient's health narrative, and how do inaccurate or failed communications among health care providers affect patient safety and outcomes?

4. Compare Madeleine to one of Poe's other female characters in this collection. Identify similarities and differences in their illness experiences and their relationships with others.

5. Research the history of the miasma theory, also known as miasmatic theory. Then locate evidence of miasmatic pollution in the story. What are we to make of the fact that the House of Usher collapses into the tarn at the end of the tale? In the twenty-first century, we know that genes, environment, and lifestyle choices play a role in health, and many health care providers are attuned to biopsychosocial factors affecting the health of their patients. Can Madeleine and Roderick's illnesses be attributed to any one source? Provide evidence to support your claims.

The Black Cat

August 1843, Philadelphia *United States Saturday Post*

In July 1842, during home renovation, a man found the bones of a young woman with a bullet hole in her skull behind a cellar wall. The bones were believed to belong to a neighborhood resident named Kendall, who had disappeared two decades before and had never been found. This chilling true-life tale of entombment may be a source for Poe's "The Black Cat," which involves an alleged alcoholic's murder and secret burial of his wife (Reilly 1993). If so, "The Black Cat" represents another example of Poe's incorporation of actual horrors into his fiction. But what inspired Poe to associate alcoholism with "perverseness"? How did his portrait of an alcoholic madman reflect scientific and moral attitudes toward alcohol abuse? And was he ahead of his time in understanding the underlying mechanisms of alcoholism?

Certainly, Poe's own struggles with alcohol must have fed his interest in the subject. Poe's personality was severely altered by even one drink, perhaps a reaction caused by some underlying pathology, such as the brain lesion with which Mary Louise Shew once diagnosed him.[1] Even knowing the devastating effects alcohol had on him, Poe still binged during difficult times, like his wife Virginia's battle with tuberculosis (Matheson 1986; Meyers 1992). Since Poe's father, David Poe, struggled with alcoholism, young Edgar may have been genetically predisposed to the same, but because he could abstain for long periods of time, his true difficulty with alcohol was temperance when drinking. When he began to drink, he could not stop and would binge for days at a time. Unfortunately, while the sober Poe was generally modest and well-mannered, when drunk, he became a "frenzied beast" (Meyers 1992, 87). One can see then how generations of readers have found in Poe the same wildness that characterizes his madman narrators, including that of "The Black Cat."

Poe's interest in alcoholism as a theme was not merely personal, however. While Poe featured drinking in a number of his stories, including "The Cask of Amontillado," "The Man of the Crowd," and "Hop-Frog," it is in "The Black Cat" that he seems to touch most self-consciously upon nineteenth-century temperance movements—and to adapt and transcend the genre that was so popular that more than 12 percent of the novels published in 1830s America included temperance themes (Matheson 1986, 70).

The temperance movement, spurred on by the religious revival called the Second Great Awakening, represented a shift in medical and moral attitudes toward the alcoholic. Dr. Benjamin Rush, the most influential American physician of the time (Sloane 1966), was one of the first to call alcoholism a disease and "advocate compassion for the drunkard" (Matheson 1986, 70). By 1838, Samuel Woodward, a prominent mental health expert, exonerated the alcoholic of moral failure, arguing that the alcoholic addiction was rooted in physiology, not morality. American temperance groups quickly picked up the medical party line that alcohol, not the alcoholic, was to blame (72). Writers, in turn, began to portray alcoholic characters with sensitivity and compassion. They also created the subgenre of "dark temperance" literature, a category into which David S. Reynolds (1988) places Poe's story. Poe added complexity to the typically predictable temperance tract, requiring the reader to analyze his narrator's motives quite carefully. Like Poe himself, the narrator may adapt the temperance tale for his own purposes—to get away with murder (Amper 1992).

On a separate note, the cat figure now has additional significance in Poe lore. As an exercise at a medical conference in 1996, cardiologist R. Michael Benitez was presented the clinical details surrounding the death of an anonymous patient. Based on those details, he concluded that his patient had died of rabies ("Poe's Death Is Rewritten" 1996). Because Poe enjoyed the company of cats and was known to have housecats and other pets at a time when animals were not inoculated against rabies, this theory is intriguing. Many Poe scholars, however, find the claim unsubstantiated (Peeples 1998; Walsh 2000), citing a flimsy understanding of Poe's biography. Nonetheless, this emphasis on biography aligns literary scholars with the health humanities, where understanding of the patient's story is necessary for understanding their illness.

•

FOR the most wild, yet most homely narrative which I am about to pen, I neither expect nor solicit belief. Mad indeed would I be to expect it, in a case where my very senses reject their own evidence. Yet, mad am I not—and very surely do I not dream. But tomorrow I die, and to-day I would unburthen my soul. My immediate purpose is to place before the world, plainly, succinctly, and without comment, a series of mere household events. In their consequences, these events have terrified—have tortured—have destroyed me. Yet I will not attempt to expound them. To me, they have presented little but Horror—to many they will seem less terrible than *baroques*. Hereafter, perhaps, some intellect may be found which will reduce my phantasm to the common-place—some intellect more calm, more logical, and far less excitable than my own, which will perceive, in the circumstances I detail with awe, nothing more than an ordinary succession of very natural causes and effects.

From my infancy I was noted for the docility and humanity of my disposition. My tenderness of heart was even so conspicuous as to make me the jest of my companions. I was especially fond of animals, and was indulged by my parents with a great variety of pets. With these I spent most of my time, and never was so happy as when feeding and caressing them. This peculiarity of character grew with my growth, and, in my manhood, I derived from it one of my principal sources of pleasure. To those who have cherished an affection for a faithful and sagacious dog, I need hardly be at the trouble of explaining the nature or the intensity of the gratification thus derivable. There is something in the unselfish and self-sacrificing love of a brute, which goes directly to the heart of him who has had frequent occasion to test the paltry friendship and gossamer fidelity of mere *Man*.

I married early, and was happy to find in my wife a disposition not uncongenial with my own. Observing my partiality for domestic pets, she lost no opportunity of procuring those of the most agreeable kind. We had birds, gold fish, a fine dog, rabbits, a small monkey, and *a cat*.

This latter was a remarkably large and beautiful animal, entirely black, and sagacious to an astonishing degree. In speaking of his intelligence, my wife, who at heart was not a little tinctured with superstition, made frequent allusion to the ancient popular notion, which regarded all black cats as witches in disguise. Not that she was ever *serious* upon this point—and I mention the matter at all for no better reason than that it happens, just now, to be remembered.

Pluto—this was the cat's name—was my favorite pet and playmate. I alone fed him, and he attended me wherever I went about the house. It was even with difficulty that I could prevent him from following me through the streets.

Our friendship lasted, in this manner, for several years, during which my general temperament and character—through the instrumentality of the Fiend Intemperance—had (I blush to confess it) experienced a radical alteration for the worse. I grew, day by day, more moody, more irritable, more regardless of the feelings of others. I suffered myself to use intemperate language to my wife. At length, I even offered her personal violence. My pets, of course, were made to feel the change in my disposition. I not only neglected, but ill-used them. For Pluto, however, I still retained sufficient regard to restrain me from maltreating him, as I made no scruple of maltreating the rabbits, the monkey, or even the dog, when, by accident, or through affection, they came in my way. But my disease grew upon me—for what disease is like Alcohol!—and at length even Pluto, who was now becoming old, and consequently somewhat peevish—even Pluto began to experience the effects of my ill temper.

One night, returning home, much intoxicated, from one of my haunts about town, I fancied that the cat avoided my presence. I seized him; when, in his fright at my violence, he inflicted a slight wound upon my hand with his teeth. The fury of a demon instantly possessed me. I knew myself no longer. My original soul seemed, at once, to take its flight from my body; and a more than fiendish malevolence, gin-nurtured, thrilled every fibre of my frame. I took from my waistcoat-pocket a pen-knife, opened it, grasped the poor beast by the throat, and deliberately cut one of its eyes from the socket! I blush, I burn, I shudder, while I pen the damnable atrocity.

When reason returned with the morning—when I had slept off the fumes of the night's debauch—I experienced a sentiment half of horror, half of remorse, for the crime of which I had been guilty; but it was, at best, a feeble and equivocal feeling, and the soul remained untouched. I again plunged into excess, and soon drowned in wine all memory of the deed.

In the meantime the cat slowly recovered. The socket of the lost eye presented, it is true, a frightful appearance, but he no longer appeared to suffer any pain. He went about the house as usual, but, as might be expected, fled in extreme terror at my approach. I had so much of my old heart left, as to be at first grieved by this evident dislike on the part of a creature which had once so loved me. But this feeling soon gave place to irritation. And then came, as if to my final and irrevocable overthrow, the spirit of PERVERSENESS. Of this spirit philosophy takes no account. Yet I am not more sure that my soul lives, than I am that perverseness is one of the primitive impulses of the human heart—one of the indivisible primary faculties, or sentiments, which give direction to the character of Man. Who has not, a hundred times, found himself committing a vile or a stupid action, for no other reason than because he knows he should *not?* Have we not a perpetual inclination, in the teeth of our best judgment, to violate that which is *Law,* merely because we understand it to be such? This spirit of perverseness, I say, came to my final overthrow. It was this unfathomable longing of the soul *to vex itself*—to offer violence to its own nature—to do wrong for the wrong's sake only—that urged me to continue and finally to consummate the injury I had inflicted upon the unoffending brute. One morning, in cool blood, I slipped a noose about its neck and hung it to the limb of a tree;—hung it with the tears streaming from my eyes, and with the bitterest remorse at my heart;—hung it *because* I knew that it had loved me, and *because* I felt it had given me no reason of offence;—hung it *because* I knew that in so doing I was committing a sin—a deadly sin that would so jeopardize my immortal soul as to place it—if such a thing were possible—even beyond the reach of the infinite mercy of the Most Merciful and Most Terrible God.

On the night of the day on which this most cruel deed was done, I was aroused from sleep by the cry of fire. The curtains of my bed were in flames. The whole house was blazing. It was with great difficulty that my wife, a servant, and myself, made our escape from the conflagration. The destruction was complete. My entire worldly wealth was swallowed up, and I resigned myself thenceforward to despair.

I am above the weakness of seeking to establish a sequence of cause and effect, between the disaster and the atrocity. But I am detailing a chain of facts—and wish not to leave even a possible link imperfect. On the day succeeding the fire, I visited the ruins. The walls, with one exception, had fallen in. This exception was found in a compartment wall, not very thick, which stood about the middle of the house, and against which had rested the head of my bed. The plastering had here, in great measure, resisted the action of fire—a fact which I attributed to its having been recently spread. About this wall a dense crowd were collected, and many persons seemed to be examining a particular portion of it with very minute and eager attention. The words "strange!" "singular!" and other similar expressions, excited my curiosity. I approached and saw, as if graven in *bas relief* upon the white surface, the figure of a gigantic *cat*. The impression was given with an accuracy truly marvellous. There was a rope about the animal's neck.

When I first beheld this apparition—for I could scarcely regard it as less—my wonder and my terror were extreme. But at length reflection came to my aid. The cat, I remembered, had been hung in a garden adjacent to the house. Upon the alarm of fire, this garden had been immediately filled by the crowd—by some one of whom the animal must have been cut from the tree and thrown, through an open window, into my chamber. This had probably been done with the view of arousing me from sleep. The falling of other walls had compressed the victim of my cruelty into the substance of the freshly-spread plaster; the lime of which, with the flames, and the *ammonia* from the carcass, had then accomplished the portraiture as I saw it.

Although I thus readily accounted to my reason, if not altogether to my conscience, for the startling fact just detailed, it did not the less fail to make a deep impression upon my fancy. For months I could not rid myself of the phantasm of the cat; and, during this period, there came back into my spirit a half-sentiment that seemed, but was not, remorse. I went so far as to regret the loss of the animal, and to look about me, among the vile haunts which I now habitually frequented, for another pet of the same species, and of somewhat similar appearance, with which to supply its place.

One night as I sat, half stupefied, in a den of more than infamy, my attention was suddenly drawn to some black object, reposing upon the head of one of the immense hogsheads of Gin, or of Rum, which constituted the chief furniture of the apartment. I had been looking steadily at the top of this hogshead for some minutes, and what now caused me surprise was the fact that I had not sooner perceived the object thereupon. I approached it, and touched it with my hand. It was a black cat—a very large one—fully as large as Pluto, and closely resembling him in every respect but one. Pluto had not a white hair upon any portion of his body; but this cat had a large, although indefinite splotch of white, covering nearly the whole region of the breast.

Upon my touching him, he immediately arose, purred loudly, rubbed against my hand, and appeared delighted with my notice. This, then, was the very creature of which I was in search. I at once offered to purchase it of the landlord; but this person made no claim to it—knew nothing of it—had never seen it before.

I continued my caresses, and, when I prepared to go home, the animal evinced a disposition to accompany me. I permitted it to do so; occasionally stooping and patting it as I proceeded. When it reached the house it domesticated itself at once, and became immediately a great favorite with my wife.

For my own part, I soon found a dislike to it arising within me. This was just the reverse of what I had anticipated; but I know not how or why it was—its evident fondness for myself rather disgusted and annoyed. By slow degrees, these feelings of disgust and annoyance rose into the bitterness of hatred. I avoided the creature; a certain sense of shame, and the remembrance of my former deed of cruelty, preventing me from physically abusing it. I did not, for some weeks, strike, or otherwise violently ill use it; but gradually—very gradually—I came to look upon it with unutterable loathing, and to flee silently from its odious presence, as from the breath of a pestilence.

What added, no doubt, to my hatred of the beast, was the discovery, on the morning after I brought it home, that, like Pluto, it also had been deprived of one of its eyes. This circumstance, however, only endeared it to my wife, who, as I have already said, possessed, in a high degree, that humanity of feeling which had once been my distinguishing trait, and the source of many of my simplest and purest pleasures.

With my aversion to this cat, however, its partiality for myself seemed to increase. It followed my footsteps with a pertinacity which it would be difficult to make the reader comprehend. Whenever I sat, it would crouch beneath my chair, or spring upon my knees, covering me with its loathsome caresses. If I arose to walk it would get between my feet and thus nearly throw me down, or, fastening its long and sharp claws in my dress, clamber, in this manner, to my breast. At such times, although I longed to destroy it with a blow, I was yet withheld from so doing, partly by a memory of my former crime, but chiefly—let me confess it at once—by absolute *dread* of the beast.

This dread was not exactly a dread of physical evil—and yet I should be at a loss how otherwise to define it. I am almost ashamed to own—yes, even in this felon's cell, I am almost ashamed to own—that the terror and horror with which the animal inspired me, had been heightened by one of the merest chimæras it would be possible to conceive. My wife had called my attention, more than once, to the character of the mark of white hair, of which I have spoken, and which constituted the sole visible difference between the strange beast and the one I had destroyed. The reader will remember that this mark, although large, had been originally very indefinite; but, by slow degrees—degrees nearly imperceptible, and which for a long time my Reason struggled to reject as fanciful—it had, at length, assumed a rigorous distinctness of outline. It was now the representa-tion of an object that I shudder to name—and for this, above all, I loathed, and dreaded, and would have rid myself of the monster *had I dared*—it was now, I say, the image of a hideous—of a ghastly thing—of the GALLOWS!—oh, mournful and terrible engine of Horror and of Crime—of Agony and of Death!

And now was I indeed wretched beyond the wretchedness of mere Humanity. And *a brute beast*—whose fellow I had contemptuously destroyed—*a brute beast* to work out for *me*—for me a man fashioned in the image of the High God—so much of insufferable wo! Alas! neither by day nor by night knew I the blessing of Rest any more! During the former the creature left me no moment alone; and, in the latter, I started, hourly, from dreams of unutterable fear to find the hot breath of *the thing* upon my face, and its vast weight—an incarnate Night-Mare that I had no power to shake off—incumbent eternally upon my *heart!*

Beneath the pressure of torments such as these, the feeble remnant of the good within me succumbed. Evil thoughts became my sole intimates—the darkest and most evil of thoughts. The moodiness of my usual temper increased to hatred of all things and of all mankind; while, from the sudden, frequent, and ungovernable outbursts of a fury to which I now blindly abandoned myself, my uncomplaining wife, alas! was the most usual and the most patient of sufferers.

One day she accompanied me, upon some household errand, into the cellar of the old building which our poverty compelled us to inhabit. The cat followed me down the steep stairs, and, nearly throwing me headlong, exasperated me to madness. Uplifting an axe, and forgetting, in my wrath, the childish dread which had hitherto stayed my hand, I aimed a blow at the animal, which, of course, would have proved instantly fatal had it descended as I wished. But this blow was arrested by the hand of my wife. Goaded, by the interference, into a rage more than demoniacal, I withdrew my arm from her grasp and buried the axe in her brain. She fell dead upon the spot without a groan.

This hideous murder accomplished, I set myself forthwith, and with entire deliberation, to the task of concealing the body. I knew that I could not remove it from the house, either by day or by night, without the risk of being observed by the neighbors. Many projects entered my mind. At one period I thought of cutting the corpse into minute fragments, and destroying them by fire. At another, I resolved to dig a grave for it in the floor of the cellar. Again, I deliberated about casting it in the well in the yard—about packing it in a box, as if merchandize, with the usual arrangements, and so getting a porter to take it from the house. Finally I hit upon what I considered a far better expedient than either of these. I determined to wall it up in the cellar—as the monks of the middle ages are recorded to have walled up their victims.

For a purpose such as this the cellar was well adapted. Its walls were loosely constructed, and had lately been plastered throughout with a rough plaster, which the dampness of the atmosphere had prevented from hardening. Moreover, in one of the walls was a projection, caused by a false chimney, or fireplace, that had been filled up and made to resemble the rest of the cellar. I made no doubt that I could readily displace the bricks at this point, insert the corpse, and wall the whole up as before, so that no eye could detect anything suspicious.

And in this calculation I was not deceived. By means of a crow-bar I easily dislodged the bricks, and, having carefully deposited the body against the inner wall, I propped it in that position, while, with little trouble, I re-laid the whole structure as it originally stood. Having procured mortar, sand, and hair, with every possible precaution, I prepared a plaster which could not be distinguished from the old, and with this I very carefully went over the new brick-work. When I had finished, I felt satisfied that all was right. The wall did not present the slightest appearance of having been disturbed. The rubbish on the floor was picked up with the minutest care. I looked around triumphantly, and said to myself—"Here at least, then, my labor has not been in vain."

My next step was to look for the beast which had been the cause of so much wretchedness; for I had, at length, firmly resolved to put it to death. Had I been able to meet with it, at the moment, there could have been no doubt of its fate; but it appeared that the crafty animal had been alarmed at the violence of my previous anger, and forbore to present itself in my present mood. It is impossible to describe, or to imagine, the deep, the blissful sense of relief which the absence of the detested creature occasioned in my bosom. It did not make its appearance during the night—and thus for one night at least, since its intro-duction into the house, I soundly and tranquilly slept; aye, *slept* even with the burden of murder upon my soul!

The second and the third day passed, and still my tormentor came not. Once again I breathed as a freeman. The monster, in terror, had fled the premises for-ever! I should behold it no more! My happiness was supreme! The guilt of my dark deed disturbed me but little. Some few inquiries had been made, but these had been readily answered. Even a search had been instituted—but of course nothing was to be discovered. I looked upon my future felicity as secured.

Upon the fourth day of the assassination, a party of the police came, very unexpectedly, into the house, and proceeded again to make rigorous investigation of the premises. Secure, however, in the inscrutability of my place of conceal-ment, I felt no embarrassment whatever. The officers bade me accompany them in their search. They left no nook or corner unexplored. At length, for the third or fourth time, they descended into the cellar. I quivered not in a muscle. My heart beat calmly as that of one who slumbers in innocence. I walked the cellar from end to end. I folded my arms upon my bosom, and roamed easily to and fro. The police were thoroughly satisfied and prepared to depart. The glee at my heart was too strong to be restrained. I burned to say if but one word, by way of triumph, and to render doubly sure their assurance of my guiltlessness.

"Gentlemen," I said at last, as the party ascended the steps, "I delight to have allayed your suspicions. I wish you all health, and a little more courtesy. By the bye, gentlemen, this—this is a very well constructed house." [In the rabid desire to say something easily, I scarcely knew what I uttered at all.]—"I may say an *ex-cellently* well constructed house. These walls—are you going, gentlemen?—these walls are solidly put together;" and here, through the mere phrenzy of bravado, I rapped heavily with a cane which I held in my hand, upon that very portion of the brick-work behind which stood the corpse of the wife of my bosom.

But may God shield and deliver me from the fangs of the Arch-Fiend! No sooner had the reverberation of my blows sunk into silence, than I was answered by a voice from within the tomb!—by a cry, at first muffled and broken, like the sobbing of a child, and then quickly swelling into one long, loud, and con-tinuous scream, utterly anomalous and inhuman—a howl—a wailing shriek, half of horror and half of triumph, such as might have arisen only out of hell, conjointly from the throats of the damned in their agony and of the demons that exult in the damnation.

Of my own thoughts it is folly to speak. Swooning, I staggered to the oppo-site wall. For one instant the party on the stairs remained motionless, through extremity of terror and awe. In the next, a dozen stout arms were toiling at the wall. It fell bodily. The corpse, already greatly decayed and clotted with gore, stood erect before the eyes of the spectators. Upon its head, with red extended mouth and solitary eye of fire, sat the hideous beast whose craft had seduced me into murder, and whose informing voice had consigned me to the hangman. I had walled the monster up within the tomb!

DISCUSSION QUESTIONS

1. The narrator suggests "perverseness" plays into his actions. Review the passage on perverseness and try to describe it to another classmate. In what ways does Poe's theory of perverseness challenge nineteenth-century American medical practices based on the theory that all illness could be understood through observation of the physical body? What explanations might medical science offer for this perverseness today?

2. One scholar (Matheson 1986) has argued that Poe's narrator is not an alcoholic but a man who appropriates the well-known temperance narrative of his time to portray himself in a more sympathetic light. In short, he uses alcohol as an excuse for his crime. Locate evidence in the story to support this reading. Then, consider the power we have when we get to control our own narrative—or the powerlessness we might experience when someone else is given the power to tell our story. What implications does this have for patients in health care?

3. What attitudes exist toward addiction today? Do you believe the average health care provider is as empathetic toward, say, a meth addict as he or she is toward a cancer patient? Why or why not? Do these different responses reflect lingering moral judgments attached to certain diseases? What methods might current or future health care providers use to examine their own biases and assumptions when serving stigmatized or marginalized populations?

4. Analyze the narrator's relationship with his pets and his wife. How do the two differ? How does the narrator's wife compare to the women who meet tragic ends in other Poe tales? Based on your readings of two or more Poe stories, do you believe Poe was trying to give voice to women, who were so often unvoiced in life and literature of the time, or did he silence and stereotype women?

5. Select a representation of addiction from another genre of media—film, photography, painting, et cetera—and write a brief reflection about what that work says to you about the experience of addiction. Compare it to the experience of addiction portrayed in "The Black Cat."

6. Although Poe scholarship has proved otherwise, Poe is still often mischaracterized as a writer who wrote to purge personal demons, like alcoholism, and

is conflated with his tormented madmen. While this view of Poe is misguided, it points to our cultural belief that writing can be therapeutic—a way to expel our demons. What research exists to support the therapeutic value of writing? Conduct a literature review addressing this question. Hint: The work of James Pennebaker is a good place to start!

A Tale of the Ragged Mountains

April 1844, *Godey's Lady's Book*

In the preface to his translation of Poe's "Mesmeric Revelation," Charles Baudelaire wrote that the ideas surrounding mesmerism—also called animal magnetism—had "entered the minds of the poets at the same time that they [had] entered the minds of scientists" (Falk 1969, 537). Like other romantic writers, Poe was keen to engage the topic, which, like phrenology, fascinated the American public. Poe's series of mesmerism stories includes "A Tale of the Ragged Mountains," "Mesmeric Revelation," and "The Facts in the Case of M. Valdemar," with each tale investigating a different aspect of the powers of mesmerism and the mesmerist-patient relationship.

"A Tale of the Ragged Mountains," the first in the series, is a case study (Lind 1947) in mesmerism in which a patient named Bedloe is treated for neuralgia by Doctor Templeton, who has established such a strong mesmeric connection with the former that he is able to put him into a trance instantaneously. This connection between doctor and patient reflects mid-nineteenth-century teachings. One mesmerist identified only as "A Physician," authored The Animal Magnetizer, *proposing that the effective use of magnetism relied upon "a moral and physical sympathy" that created "perfect communication" between magnetizer and patient, even across great distances (as quoted in Lind 1947, 1082). Charles P. Johnson, another mesmerist, stresses in his 1844* A Treatise of Animal Magnetism *that it is imperative for the mesmerist to keep the patient's welfare first in his mind: "To act efficaciously, he should feel himself drawn toward the person who requires his care, take interest in him, and have the desire and hope of curing, or at least relieving him. As soon as he has decided, which he should never do lightly, he ought to consider whom he Magnetises as his brother—as his friend; he should be devoted to him as not to perceive the sacrifices that he imposes upon himself. Any other consideration, any other motive than the desire of doing good, ought not to induce him to undertake a treatment" (Johnson 1844, 18). The focus on mesmerism as a relationship-centered act of healing is intriguing. Does Poe present the mesmerist in this favorable light? Are we to read Doctor Templeton as*

a physician who partners with his patient for the patient's benefit? Or is he, like Westervelt in Hawthorne's The Blithedale Romance, *a mesmerist who enters the consciousness of another to "exert his will there" (Stoehr 1974, 386)?*

Doctor Templeton's patient, Bedloe, is also of keen interest separate from the mesmerism question. Robert Battle (2011) sees in Bedloe "a profound phenotype of Marfan syndrome" (148), another example of Poe's ability to describe physical and mental diseases long before they were fully understood or named. Poe, of course, identifies the disease that plagues Bedloe as neuralgia. Are these two diagnoses compatible? Just as Poe arguably played diagnostician in a number of his tales, readers might use their own combination of evidence and insight to find answers to these questions.

•

DURING the fall of the year 1827, while residing near Charlottesville, Virginia, I casually made the acquaintance of Mr. Augustus Bedloe. This young gentleman was remarkable in every respect, and excited in me a profound interest and curiosity. I found it impossible to comprehend him either in his moral or his physical relations. Of his family I could obtain no satisfactory account. Whence he came, I never ascertained. Even about his age—although I call him a young gentleman—there was something which perplexed me in no little degree. He certainly *seemed* young—and he made a point of speaking about his youth—yet there were moments when I should have had little trouble in imagining him a hundred years of age. But in no regard was he more peculiar than in his personal appearance. He was singularly tall and thin. He stooped much. His limbs were exceedingly long and emaciated. His forehead was broad and low. His complexion was absolutely bloodless. His mouth was large and flexible, and his teeth were more wildly uneven, although sound, than I had ever before seen teeth in a human head. The expression of his smile, however, was by no means unpleasing, as might be supposed; but it had no variation whatever. It was one of profound melancholy—of a phaseless and unceasing gloom. His eyes were abnormally large, and round like those of a cat. The pupils, too, upon any accession or diminution of light, underwent contraction or dilation, just such as is observed in the feline tribe. In moments of excitement the orbs grew bright to a degree almost inconceivable; seeming to emit luminous rays, not of a reflected but of an intrinsic lustre, as does a candle or the sun; yet their ordinary condition was so totally vapid, filmy and dull, as to convey the idea of the eyes of a long-interred corpse.

These peculiarities of person appeared to cause him much annoyance, and he was continually alluding to them in a sort of half explanatory, half apologetic strain, which, when I first heard it, impressed me very painfully. I soon, however, grew accustomed to it, and my uneasiness wore off. It seemed to be his design rather to insinuate than directly to assert that, physically, he had not always been what he was—that a long series of neuralgic attacks had reduced him from a condition of more than usual personal beauty, to that which I saw. For many years past he had been attended by a physician, named Templeton—an old gentleman, perhaps seventy years of age—whom he had first encountered at Saratoga, and from whose attention, while there, he either received, or fancied that he received, great benefit. The result was that Bedloe, who was wealthy, had made an arrangement with Dr. Templeton, by which the latter, in consideration of a liberal annual allowance, had consented to devote his time and medical experience exclusively to the care of the invalid.

Doctor Templeton had been a traveller in his younger days, and at Paris, had become a convert, in great measure, to the doctrines of Mesmer. It was altogether by means of magnetic remedies that he had succeeded in alleviating the acute pains of his patient; and this success had very naturally inspired the latter with a certain degree of confidence in the opinions from which the remedies had been educed. The Doctor, however, like all enthusiasts, had struggled hard to make a thorough convert of his pupil, and finally so far gained his point as to induce the sufferer to submit to numerous experiments.—By a frequent repetition of these, a result had arisen, which of late days has become so common as to attract little or no attention, but which, at the period of which I write, had very rarely been known in America. I mean to say, that between Doctor Templeton and Bedloe there had grown up, little by little, a very distinct and strongly marked *rapport*, or magnetic relation. I am not prepared to assert, however, that this *rapport* extended beyond the limits of the simple sleep-producing power; but this power itself had attained great intensity. At the first attempt to induce the magnetic somnolency, the mesmerist entirely failed. In the fifth or sixth he succeeded very partially, and after long continued effort. Only at the twelfth was the triumph complete. After this the will of the patient succumbed rapidly to that of the physician, so that, when I first became acquainted with the two, sleep was brought about almost instantaneously by the mere volition of the operator, even when the invalid was unaware of his presence. It is only now, in the year 1845, when similar miracles are witnessed daily by thousands, that I dare venture to record this apparent impossibility as a matter of serious fact.

The temperament of Bedloe was, in the highest degree sensitive, excitable, enthusiastic. His imagination was singularly vigorous and creative; and no doubt it derived additional force from the habitual use of morphine, which he swallowed in great quantity, and without which he would have found it impossible to exist. It was his practice to take a very large dose of it immediately after breakfast, each morning—or rather immediately after a cup of strong coffee, for he ate nothing in the forenoon—and then set forth alone, or attended only by a dog, upon a long ramble among the chain of wild and dreary hills that lie westward and southward of Charlottesville, and are there dignified by the title of the Ragged Mountains.

Upon a dim, warm, misty day, toward the close of November, and during the strange *interregnum* of the seasons which in America is termed the Indian Summer, Mr. Bedloe departed as usual, for the hills. The day passed, and still he did not return.

About eight o'clock at night, having become seriously alarmed at his protract-ed absence, we were about setting out in search of him, when he unexpectedly made his appearance, in health no worse than usual, and in rather more than ordinary spirits. The account which he gave of his expedition, and of the events which had detained him, was a singular one indeed.

"You will remember," said he, "that it was about nine in the morning when I left Charlottesville. I bent my steps immediately to the mountains, and, about ten, entered a gorge which was entirely new to me. I followed the windings of this pass with much interest.—The scenery which presented itself on all sides, although scarcely entitled to be called grand, had about it an indescribable and to me a delicious aspect of dreary desolation. The solitude seemed absolutely virgin. I could not help believing that the green sods and the gray rocks upon which I trod had been trodden never before by the foot of a human being. So entirely secluded, and in fact inaccessible, except through a series of accidents, is the entrance of the ravine, that it is by no means impossible that I was indeed the first adventurer—the very first and sole adventurer who had ever penetrated its recesses.

"The thick and peculiar mist, or smoke, which distinguishes the Indian Summer, and which now hung heavily over all objects, served, no doubt, to deepen the vague impressions which these objects created. So dense was this pleasant fog, that I could at no time see more than a dozen yards of the path before me. This path was excessively sinuous, and as the sun could not be seen, I soon lost all idea of the direction in which I journeyed. In the meantime the morphine had its customary effect—that of enduing all the external world with an intensity of interest. In the quivering of a leaf—in the hue of a blade of grass—in the shape of a trefoil—in the humming of a bee—in the gleaming of a dew-drop—in the breathing of the wind—in the faint odors that came from the forest—there came a whole universe of suggestion—a gay and motley train of rhapsodical and immethodical thought.

"Busied in this, I walked on for several hours, during which the mist deep-ened around me to so great an extent, that at length I was reduced to an absolute groping of the way. And now an indescribable uneasiness possessed me—a species of nervous hesitation and tremor.—I feared to tread, lest I should be precipitated into some abyss. I remembered, too, strange stories told about these Ragged Hills, and of the uncouth and fierce races of men who tenanted their groves and caverns. A thousand vague fancies oppressed and disconcerted me—fancies the more distressing because vague. Very suddenly my attention was arrested by the loud beating of a drum.

"My amazement was, of course, extreme. A drum in these hills was a thing unknown. I could not have been more surprised at the sound of the trump of the Archangel. But a new and still more astounding source of interest and perplexity arose. There came a wild rattling or jingling sound, as if of a bunch of large keys—and upon the instant a dusky-visaged and half-naked man rushed past me with a shriek. He came so close to my person that I felt his hot breath upon my face. He bore in one hand an instrument composed of an assemblage of steel rings, and shook them vigorously as he ran. Scarcely had he disappeared in the mist, before, panting after him, with open mouth and glaring eyes, there darted a huge beast. I could not be mistaken in its character. It was a hyena.

"The sight of this monster rather relieved than heightened my terrors—for I now made sure that I dreamed, and endeavored to arouse myself to waking consciousness. I stepped boldly and briskly forward. I rubbed my eyes. I called aloud. I pinched my limbs. A small spring of water presented itself to my view, and here, stooping, I bathed my hands and my head and neck. This seemed to dissipate the equivocal sensations which had hitherto annoyed me. I arose, as I thought, a new man, and proceeded steadily and complacently on my unknown way.

"At length, quite overcome by exertion, and by a certain oppressive closeness of the atmosphere, I seated myself beneath a tree. Presently there came a feeble gleam of sunshine, and the shadow of the leaves of the tree fell faintly but definitely upon the grass. At this shadow I gazed wonderingly for many minutes. Its character stupefied me with astonishment. I looked upward. The tree was a palm.

"I now arose hurriedly, and in a state of fearful agitation—for the fancy that I dreamed would serve me no longer. I saw—I felt that I had perfect command of my senses—and these senses now brought to my soul a world of novel and singular sensation. The heat became all at once intolerable. A strange odor loaded the breeze.—A low, continuous murmur, like that arising from a full, but gently-flowing river, came to my ears, intermingled with the peculiar hum of multitudinous human voices.

"While I listened in an extremity of astonishment which I need not attempt to describe, a strong and brief gust of wind bore off the incumbent fog as if by the wand of an enchanter.

"I found myself at the foot of a high mountain, and looking down into a vast plain, through which wound a majestic river. On the margin of this river stood an Eastern-looking city, such as we read of in the Arabian Tales, but of a character even more singular than any there described. From my position, which was far above the level of the town, I could perceive its every nook and corner, as if delineated on a map. The streets seemed innumerable, and crossed each other irregularly in all directions, but were rather long winding alleys than streets, and absolutely swarmed with inhabitants. The houses were wildly picturesque. On every hand was a wilderness of balconies, of verandas, of minarets, of shrines, and fantastically carved oriels. Bazaars abounded; and there were displayed rich wares in infinite variety and profusion—silks, muslins, the most dazzling cutlery, the most magnificent jewels and gems. Besides these things, were seen, on all sides, banners and palanquins, litters with stately dames close veiled, elephants gorgeously caparisoned, idols grotesquely hewn, drums, banners, and gongs, spears, silver and gilded maces. And amid the crowd, and the clamor, and the general intricacy and confusion—amid the million of black and yellow men, turbaned and robed, and of flowing beard, there roamed a countless multitude of holy filleted bulls, while vast legions of the filthy but sacred ape clambered, chattering and shrieking, about the cornices of the mosques, or clung to the minarets and oriels. From the swarming streets to the banks of the river, there descended innumerable flights of steps leading to bathing places, while the river itself seemed to force a passage with difficulty through the vast fleets of deeply-burthened ships that far and wide encountered its surface. Beyond the limits of the city arose, in frequent majestic groups, the palm and the cocoa, with other gigantic and weird trees of vast age; and here and there might be seen a field of rice, the thatched hut of a peasant, a tank, a stray temple, a gypsy camp, or a solitary graceful maiden taking her way, with a pitcher upon her head, to the banks of the magnificent river.

"You will say now, of course, that I dreamed; but not so. What I saw—what I heard—what I felt—what I thought—had about it nothing of the unmistakable idiosyncrasy of the dream. All was rigorously self-consistent. At first, doubting that I was really awake, I entered into a series of tests, which soon convinced me that I really was. Now, when one dreams, and, in the dream, suspects that he dreams, the suspicion *never fails to confirm itself*, and the sleeper is almost immediately aroused.—Thus Novalis errs not in saying that 'we are near waking when we dream that we dream.' Had the vision occurred to me as I describe it, without my suspecting it as a dream, then a dream it might absolutely have been, but, occurring as it did, and suspected and tested as it was, I am forced to class it among other phenomena."

"In this I am not sure that you are wrong," observed Dr. Templeton, "but proceed. You arose and descended into the city."

"I arose," continued Bedloe, regarding the Doctor with an air of profound astonishment, "I arose, as you say, and descended into the city. On my way, I fell in with an immense populace, crowding, through every avenue, all in the same direction, and exhibiting in every action the wildest excitement. Very suddenly, and by some inconceivable impulse, I became intensely imbued with personal interest in what was going on. I seemed to feel that I had an important part to play, without exactly understanding what it was. Against the crowd which environed me, however, I experienced a deep sentiment of animosity. I shrank from amid them, and, swiftly, by a circuitous path, reached and entered the city. Here all was the wildest tumult and contention. A small party of men, clad in garments half-Indian, half European, and officered by gentlemen in a uniform partly British, were engaged, at great odds, with the swarming rabble of the alleys. I joined the weaker party, arming myself with the weapons of a fallen officer, and fighting I knew not whom with the nervous ferocity of despair. We were soon overpowered by numbers, and driven to seek refuge in a species of kiosk. Here we barricaded ourselves, and, for the present, were secure. From a loop-hole near the summit of the kiosk, I perceived a vast crowd, in furious agitation, surrounding and assaulting a gay palace that overhung the river. Presently, from an upper window of this palace, there descended an effeminate-looking person, by means of a string made of the turbans of his attendants. A boat was at hand, in which he escaped to the opposite bank of the river.

"And now a new object took possession of my soul. I spoke a few hurried but energetic words to my companions, and, having succeeded in gaining over a few of them to my purpose, made a frantic sally from the kiosk. We rushed amid the crowd that surrounded it. They retreated, at first, before us. They rallied, fought madly, and retreated again. In the mean time we were borne far from the kiosk, and became bewildered and entangled among the narrow streets of tall, overhanging houses, into the recesses of which the sun had never been able to shine. The rabble pressed impetuously upon us, harassing us with their spears, and overwhelming us with flights of arrows. These latter were very remarkable, and resembled in some respects the writhing creese of the Malay. They were made to imitate the body of a creeping serpent, and were long and black, with a poisoned barb. One of them struck me upon the right temple. I reeled and fell. An instantaneous and dreadful sickness seized me. I struggled—I gasped—I died."

"You will hardly persist *now*," said I, smiling, "that the whole of your adventure was not a dream. You are not prepared to maintain that you are dead?"

When I said these words, I of course expected some lively sally from Bedloe in reply; but, to my astonishment, he hesitated, trembled, became fearfully pallid, and remained silent. I looked toward Templeton. He sat erect and rigid in his chair—his teeth chattered, and his eyes were starting from their sockets. "Proceed!" he at length said hoarsely to Bedloe.

"For many minutes," continued the latter, "my sole sentiment—my sole feeling—was that of darkness and nonentity, with the consciousness of death. At length, there seemed to pass a violent and sudden shock through my soul, as if of electricity. With it came the sense of elasticity and of light. This latter I felt—not saw. In an instant I seemed to rise from the ground. But I had no bodily, no visible, audible, or palpable presence. The crowd had departed. The tumult had ceased. The city was in comparative repose. Beneath me lay my corpse, with the arrow in my temple, the whole head greatly swollen and disfigured. But all these things I felt—not saw. I took interest in nothing. Even the corpse seemed a matter in which I had no concern. Volition I had none, but appeared to be impelled into motion, and flitted buoyantly out of the city, retracing the circuitous path by which I had entered it. When I had attained that point of the ravine in the mountains at which I had encountered the hyena, I again experienced a shock as of a galvanic battery; the sense of weight, of volition, of substance, returned. I became my original self, and bent my steps eagerly homewards—but the past had not lost the vividness of the real—and not now, even for an instant, can I compel my understanding to regard it as a dream."

"Nor was it," said Templeton, with an air of deep solemnity, "yet it would be difficult to say how otherwise it should be termed. Let us suppose only, that the soul of the man of to-day is upon the verge of some stupendous psychal discoveries. Let us content ourselves with this supposition. For the rest I have some explanation to make. Here is a water-colour drawing, which I should have shown you before, but which an unaccountable sentiment of horror has hitherto prevented me from showing."

We looked at the picture which he presented. I saw nothing in it of an extraordinary character; but its effect upon Bedloe was prodigious. He nearly fainted as he gazed. And yet it was but a miniature portrait—a miraculously accurate one, to be sure—of his own very remarkable features. At least this was my thought as I regarded it.

"You will perceive," said Templeton, "the date of this picture—it is here, scarcely visible, in this corner—1780. In this year was the portrait taken. It is the likeness of a dead friend—a Mr. Oldeb—to whom I became much attached at Calcutta, during the administration of Warren Hastings. I was then only twenty years old.—When I first saw you, Mr. Bedloe, at Saratoga, it was the miraculous similarity which existed between yourself and the painting which induced me to accost you, to seek your friendship, and to bring about those arrangements which resulted in my becoming your constant companion. In accomplishing this point, I was urged partly, and perhaps principally, by a regretful memory of the deceased, but also, in part by an uneasy, and not altogether horrorless curiosity respecting yourself.

"In your detail of the vision which presented itself to you amid the hills, you have described, with the minutest accuracy, the Indian city of Benares, upon the Holy River. The riots, the combats, the massacre, were the actual events of the insurrection of Cheyte Sing, which took place in 1780, when Hastings was put in imminent peril of his life. The man escaping by the string of turbans was Cheyte Sing himself. The party in the kiosk were sepoys and British officers, headed by Hastings. Of this party I was one, and did all I could to prevent the rash and fatal sally of the officer who fell, in the crowded alleys, by the poisoned arrow of a Bengalee. That officer was my dearest friend. It was Oldeb. You will perceive by these manuscripts," (here the speaker produced a note-book in which several pages appeared to have been freshly written) "that at the very period in which you fancied these things amid the hills, I was engaged in detailing them upon paper here at home."

In about a week after this conversation, the following paragraphs appeared in a Charlottesville paper:

"We have the painful duty of announcing the death of Mr. AUGUSTUS BEDLO, a gentleman whose amiable manners and many virtues have long endeared him to the citizens of Charlottesville.

"Mr. B., for some years past, has been subject to neuralgia, which has often threatened to terminate fatally; but this can be regarded only as the mediate cause of his decease. The proximate cause was one of especial singularity. In an excursion to the Ragged Mountains, a few days since, a slight cold and fever were contracted, attended with great determination of blood to the head. To relieve this, Dr. Templeton resorted to topical bleeding. Leeches were applied to the temples. In a fearfully brief period the patient died, when it appeared that, in the jar containing the leeches, had been, introduced, by accident, one of the venomous vermicular sangsues which are now and then found in the neighboring ponds. This creature fastened itself upon a small artery in the right temple. Its close resemblance to the medicinal leech caused the mistake to be overlooked until too late.

"N.B. The poisonous sangsue of Charlottesville may always be distinguished from the medicinal leech by its blackness, and especially by its writhing or vermicular motions, which very nearly resemble those of a snake."

I was speaking with the editor of the paper in question, upon the topic of this remarkable accident, when it occurred to me to ask how it happened that the name of the deceased had been given as Bedlo.

"I presume," said I, "you have authority for this spelling, but I have always supposed the name to be written with an *e* at the end."

"Authority?—no," he replied. "It is a mere typographical error. The name is Bedloe with an *e*, all the world over, and I never knew it to be spelt otherwise in my life."

"Then," said I mutteringly, as I turned upon my heel, "then indeed has it come to pass that one truth is stranger than any fiction—for Bedlo, without the *e*, what is it but Oldeb conversed? And this man tells me it is a typographical error."

DISCUSSION QUESTIONS

1. Why does Poe choose to have a narrator other than Bedloe or Templeton relate this tale of mesmerism? How might the story change if it were told from one of their perspectives? In what ways is it important for health care providers to possess perspective-taking skills? Can these skills, which we associate with empathy and compassion, truly make a difference in patient healing?

2. As noted in the introduction to this tale, nineteenth-century mesmerists argued that the effective use of magnetism relied upon "a moral and physical sympathy" that created "perfect communication" between magnetizer and patient (as quoted in Lind 1947, 1082). Characterize the relationship between Bedloe and Doctor Templeton. Does it meet these standards? Reflect in writing on your own relationships with your health care providers. Or, if you are a health care provider, reflect on the methods you use to establish positive rapport with patients.

3. Doctor Templeton is a believer in metempsychosis—the transmigration of the soul, or reincarnation. Do we have enough evidence to believe that a case of transmigration occurs in the story? Why or why not? Those deeply invested in further discussion of this mystery might refer to Doris Falk's 1969 article "Poe and the Power of Animal Magnetism."

4. Is Doctor Templeton responsible for Bedloe's death? Why or why not? What significance do you assign to his medical error? What methods do providers use to protect against medical error—or to minimize its effects? In what ways can effective patient-provider communication and provider-provider communication facilitate patient safety and satisfaction?

5. Research Marfan syndrome. Do you agree with physician Robert Battle that Bedloe could suffer from this disease? In Poe's time, of course, the syndrome had yet to be discovered. What do you think of Poe's seemingly uncanny ability to describe diseases before the doctors of his time did? What does this say about the contribution artists can make to scientific and medical knowledge? In what ways are artists trained to see the world differently than health care practitioners and biomedical scientists?

The Premature Burial

July 1844, Philadelphia *Dollar Newspaper*

In the winter of 2014, a seventy-eight-year-old Mississippi man, Walter Williams, woke up in a body bag, having been declared dead and loaded into the bag by the coroner and Williams's sons (Ford 2014). He was rushed into care, where he was diagnosed as alive; he lived two more weeks before meeting his final demise. Doctors suspected Williams's pacemaker stopped and then restarted, creating the appearance that he was without a heartbeat and therefore dead. The story made national news, demonstrating how rare it is today for people to be falsely declared dead.

Unfortunately, the same cannot be said of the nineteenth century. Premature burial "evoked powerful and legitimate fears in Poe's day" (Kennedy 1977, 165), as it was a reality in times when physicians did not always have reliable means of determining the line between life and death—and when bodies were often buried hastily to avoid the spread of epidemic diseases. Sometimes when graves were opened days after burials, those doing the unearthing would find evidence that the now truly deceased had attempted to escape: torn clothes, shredded fingernails, scratches on the inside of the coffin, even broken limbs. Many in the medical profession claimed cases of premature burial were exaggerated, but in 1819, a French physician wrote that as many as one-third of dying patients were buried prematurely (Dibble 2010).

Premature burial was considered such an important problem that one Baltimore man invented a "life-preserving coffin" that allowed the buried individual to pull on a rope connected to a bell above ground, alerting the living of their devastating mistake (Kennedy 1977, 165). A British man published an influential book called Premature Burial, and How It May Be Prevented *(Dibble 2010). In America, a wide variety of narratives, from "quasi-medical reports" to "hysterical recollections" (Kennedy 1977, 160), stirred American's fears. Familiar with the premature burial genre and not one to pass up the opportunity to capitalize on a sensational topic, Poe, of course, tried his own hand at the theme. In fact, premature burial is a motif in a number of his tales, but it is in "The Premature Burial" that he tackles the theme most directly.*

While at least one scholar has argued compellingly that Poe's narrator could be diagnosed as phobic and paranoid by today's diagnostic criteria (Zimmerman 2009), we might also argue that although the narrator is expressing personal fears he is, more importantly, reflecting cultural phobias. In short, in this tale, as in others in this collection, including "Berenice," "The Fall of the House of Usher," and "Some Words with a Mummy," Poe underscores a cultural paranoia that accompanied the rise of empirical medicine—the suspicion that some elements of the medical community found individuals more valuable in death, when the body could be dissected and studied. Certain factions of the medical community even accused other groups of physicians of being more invested in conducting postmortem exams than in curing patients (Browner 2005).

Yet, while it's tempting to believe Poe's tale is strictly a horror story of premature burial, we cannot overlook that the tale is also a hoax underpinned by dark humor. By combining various genres of narrative within the tale and blurring the line between fiction and nonfiction, horror and humor, Poe prompts us to remember that narrative interpretation is a complex endeavor. We are reminded that having the skills to interpret a narrative—while remaining open to its ambiguity—is of particular importance in health care today. The tale itself provides good practice for developing those skills.

•

THERE are certain themes of which the interest is all-absorbing, but which are too entirely horrible for the purposes of legitimate fiction. These the mere romanticist must eschew, if he do not wish to offend, or to disgust. They are with propriety handled, only when the severity and majesty of Truth sanctify and sustain them. We thrill, for example, with the most intense of "pleasurable pain," over the accounts of the Passage of the Beresina, of the Earthquake at Lisbon, of the Plague at London, of the Massacre of St. Bartholomew, or of the stifling of the hundred and twenty-three prisoners in the Black Hole at Calcutta. But, in these accounts, it is the fact—it is the reality—it is the history which excites. As inventions, we should regard them with simple abhorrence.

I have mentioned some few of the more prominent and august calamities on record; but, in these, it is the extent, not less than the character of the calamity, which so vividly impresses the fancy. I need not remind the reader that, from the long and weird catalogue of human miseries, I might have selected many individual instances more replete with essential suffering than any of these vast generalities of disaster. The true wretchedness, indeed—the ultimate woe—is particular, not diffuse. That the ghastly extremes of agony are endured by man the unit, and never by man the mass—for this let us thank a merciful God!

To be buried while alive, is, beyond question, the most terrific of these extremes which has ever fallen to the lot of mere mortality. That it has frequently, very frequently, so fallen, will scarcely be denied by those who think. The boundaries which divide Life from Death, are at best shadowy and vague. Who shall say where the one ends, and where the other begins? We know that there are diseases in which occur total cessations of all the apparent functions of vitality, and yet in which these cessations are merely suspensions, properly so called. They are only temporary pauses in the incomprehensible mechanism. A certain period elapses, and some unseen mysterious principle again sets in motion the magic pinions and the wizard wheels. The silver cord was not for ever loosed, nor the golden bowl irreparably broken. But where, meantime, was the soul?

Apart, however, from the inevitable conclusion, *a priori,* that such causes must produce such effects—that the well known occurrence of such cases of suspended animation must naturally give rise, now and then, to premature interments—apart from this consideration, we have the direct testimony of medical and ordinary experience, to prove that a vast number of such interments have actually taken place. I might refer at once, if necessary, to a hundred well authenticated instances. One of very remarkable character, and of which the circumstances may be fresh in the memory of some of my readers, occurred, not very long ago, in the neighboring city of Baltimore, where it occasioned a painful, intense, and widely extended excitement. The wife of one of the most respectable citizens—a lawyer of eminence and a member of Congress—was seized with a sudden and unaccountable illness, which completely baffled the skill of her physicians. After much suffering she died, or was supposed to die. No one suspected, indeed, or had reason to suspect, that she was not actually dead. She presented all the ordinary appearances of death. The face assumed the usual pinched and sunken outline. The lips were of the usual marble pallor. The eyes were lustreless. There was no warmth. Pulsation had ceased. For three days the body was preserved unburied, during which it had acquired a stony rigidity. The funeral, in short, was hastened, on account of the rapid advance of what was supposed to be decomposition.

The lady was deposited in her family vault, which, for three subsequent years, was undisturbed. At the expiration of this term, it was opened for the reception of a sarcophagus;—but, alas! how fearful a shock awaited the husband, who, personally, threw open the door. As its portals swung outwardly back, some white-apparelled object fell rattling within his arms. It was the skeleton of his wife in her yet unmouldered shroud.

A careful investigation rendered it evident that she had revived within two days after her entombment—that her struggles within the coffin had caused it to fall from a ledge, or shelf, to the floor, where it was so broken as to permit her escape. A lamp which had been accidentally left, full of oil, within the tomb, was found empty; it might have been exhausted, however, by evaporation. On the uppermost of the steps which led down into the dread chamber, was a large fragment of the coffin, with which, it seemed that she had endeavored to arrest attention, by striking the iron door. While thus occupied, she probably swooned, or possibly died, through sheer terror; and, in falling, her shroud became entangled in some iron-work which projected interiorly. Thus she remained, and thus she rotted, erect.

In the year 1810, a case of living inhumation happened in France, attended with circumstances which go far to warrant the assertion that truth is, indeed, stranger than fiction. The heroine of the story was a Mademoiselle Victorine Lafourcade, a young girl of illustrious family, of wealth, and of great personal beauty. Among her numerous suitors was Julien Bossuet, a poor *littérateur,* or journalist, of Paris. His talents and general amiability had recommended him to the notice of the heiress, by whom he seems to have been truly beloved; but her pride of birth decided her, finally, to reject him, and to wed a Monsieur Rénelle, a banker, and a diplomatist of some eminence. After marriage, however, this gentleman neglected, and, perhaps, even more positively ill-treated her. Having passed with him some wretched years, she died,—at least her condition so closely resembled death as to deceive every one who saw her. She was buried—not in a vault—but in an ordinary grave in the village of her nativity. Filled with despair, and still inflamed by the memory of a profound attachment, the lover journeys from the capital to the remote province in which the village lies, with the romantic purpose of disinterring the corpse, and possessing himself of its luxuriant tresses. He reaches the grave. At midnight he unearths the coffin, opens it, and is in the act of detaching the hair, when he is arrested by the unclosing of the beloved eyes. In fact, the lady had been buried alive. Vitality had not altogether departed; and she was aroused, by the caresses of her lover, from the lethargy which had been mistaken for death. He bore her frantically to his lodgings in the village. He employed certain powerful restoratives suggested by no little medical learning. In fine, she revived. She recognized her preserver. She remained with him until, by slow degrees, she fully recovered her original health. Her woman's heart was not adamant, and this last lesson of love sufficed to soften it. She bestowed it upon Bossuet. She returned no more to her husband, but concealing from him her resurrection, fled with her lover to America. Twenty years afterward, the two returned to France, in the persuasion that time had so greatly altered the lady's appearance that her friends would be unable to recognize her. They were mistaken, however; for, at the first meeting, Monsieur Rénelle did actually recognize and make claim to his wife. This claim she resisted; and a judicial tribunal sustained her in her resistance; deciding that the peculiar circumstances, with the long lapse of years, had extinguished, not only equitably but legally, the authority of the husband.

The "Chirurgical Journal" of Leipsic—a periodical, of high authority and merit, which some American bookseller would do well to translate and re-publish—records, in a late number, a very distressing event of the character in question.

An officer of artillery, a man of gigantic stature and of robust health, being thrown from an unmanageable horse, received a very severe contusion upon the head, which rendered him insensible at once; the skull was slightly frac-tured; but no immediate danger was apprehended. Trepanning was accom-plished successfully. He was bled, and many other of the ordinary means of relief were adopted. Gradually, however, he fell into a more and more hopeless state of stupor; and, finally, it was thought that he died.

The weather was warm; and he was buried, with indecent haste, in one of the public cemeteries. His funeral took place on Thursday. On the Sunday following, the grounds of the cemetery were, as usual, much thronged with visiters; and, about noon, an intense excitement was created by the declaration of a peasant, that, while sitting upon the grave of the officer, he had distinctly felt a commotion of the earth, as if occasioned by some one struggling beneath. At first little attention was paid to the man's asseveration; but his evident terror, and the dogged obstinacy with which he persisted in his story, had, at length, their natural effect upon the crowd. Spades were hurriedly procured, and the grave, which was shamefully shallow, was, in a few minutes, so far thrown open that the head of its occupant appeared. He was then, seemingly, dead; but he sat nearly erect within his coffin, the lid of which, in his furious struggles, he had partially uplifted.

He was forthwith conveyed to the nearest Hospital, and there pronounced to be still living, although in an asphytic condition. After some hours he revived, recognized individuals of his acquaintance, and, in broken sentences, spoke of his agonies in the grave.

From what he related, it was clear that he must have been conscious of life for more than an hour, while inhumed, before lapsing into insensibility. The grave was carelessly and loosely filled with an exceedingly porous soil; and thus some air was necessarily admitted. He heard the footsteps of the crowd overhead, and endeavored to make himself heard in turn. It was the tumult within the grounds of the cemetery, he said, which appeared to awaken him from a deep sleep—but no sooner was he awake than he became fully aware of the awful horrors of his position.

This patient, it is recorded, was doing well, and seemed to be in a fair way of ultimate recovery, but fell a victim to the quackeries of medical experiment. The galvanic battery was applied; and he suddenly expired in one of those ecstatic paroxysms which, occasionally, it superinduces.

The mention of the galvanic battery, nevertheless, recalls to my memory a well known and very extraordinary case in point, where its action proved the means of restoring to animation a young attorney of London who had been interred for two days. This occurred in 1831, and created, at the time, a very profound sensation wherever it was made the subject of converse.

The patient, Mr. Edward Stapleton, had died, apparently, of typhus fever, accompanied with some anomalous symptoms which had excited the curiosity of his medical attendants. Upon his seeming decease, his friends were requested to sanction a *post mortem* examination, but declined to permit it. As often happens, when such refusals are made, the practitioners resolved to disinter the body and dissect it at leisure, in private. Arrangements were easily effected with some of the numerous corps of body-snatchers with which London abounds; and, upon the third night after the funeral, the supposed corpse was unearthed from a grave eight feet deep, and deposited in the operating chamber of one of the private hospitals.

An incision of some extent had been actually made in the abdomen, when the fresh and undecayed appearance of the subject suggested an application of the battery. One experiment succeeded another, and the customary effects supervened, with nothing to characterize them in any respect, except, upon one or two occasions, a more than ordinary degree of life-likeness in the convulsive action.

It grew late. The day was about to dawn; and it was thought expedient, at length, to proceed at once to the dissection. A student, however, was especially desirous of testing a theory of his own, and insisted upon applying the battery to one of the pectoral muscles. A rough gash was made, and a wire hastily brought in contact; when the patient, with a hurried, but quite unconvulsive movement, arose from the table, stepped into the middle of the floor, gazed about him uneasily for a few seconds, and then—spoke. What he said was unintelligible; but words were uttered; the syllabification was distinct. Having spoken, he fell heavily to the floor.

For some moments all were paralyzed with awe—but the urgency of the case soon restored them their presence of mind. It was seen that Mr. Stapleton was alive, although in a swoon. Upon exhibition of ether he revived and was rapidly restored to health, and to the society of his friends—from whom, however, all knowledge of his resuscitation was withheld, until a relapse was no longer to be apprehended. Their wonder—their rapturous astonishment—may be conceived.

The most thrilling peculiarity of this incident, nevertheless, is involved in what Mr. S. himself asserts. He declares that at no period was he altogether insensible—that, dully and confusedly, he was aware of every thing which happened to him, from the moment in which he was pronounced *dead* by his physicians, to that in which he fell swooning to the floor of the Hospital. "I am alive," were the uncomprehended words which, upon recognizing the locality of the dissecting-room, he had endeavored, in his extremity, to utter.

It were an easy matter to multiply such histories as these—but I forbear—for, indeed, we have no need of such to establish the fact that premature interments occur. When we reflect how very rarely, from the nature of the case, we have it in our power to detect them, we must admit that they may *frequently* occur without our cognizance. Scarcely, in truth, is a graveyard ever encroached upon, for any purpose, to any great extent, that skeletons are not found in postures which suggest the most fearful of suspicions.

Fearful indeed the suspicion—but more fearful the doom! It may be asserted, without hesitation, that *no* event is so terribly well adapted to inspire the supreme-ness of bodily and of mental distress, as is burial before death. The unendurable oppression of the lungs—the stifling fumes of the damp earth—the clinging to the death garments—the rigid embrace of the narrow house—the blackness of the absolute Night—the silence like a sea that overwhelms—the unseen but palpable presence of the Conqueror Worm—these things, with the thoughts of the air and grass above, with memory of dear friends who would fly to save us if but informed of our fate, and with consciousness that of this fate they can *never* be informed—that our hopeless portion is that of the really dead—these consid-erations, I say, carry into the heart, which still palpitates, a degree of appalling and intolerable horror from which the most daring imagination must recoil. We know of nothing so agonizing upon Earth—we can dream of nothing half so hideous in the realms of the nethermost Hell. And thus all narratives upon this topic have an interest profound; an interest, nevertheless, which, through the sacred awe of the topic itself, very properly and very peculiarly depends upon our conviction of the *truth* of the matter narrated. What I have now to tell is of my own actual knowledge—of my own positive and personal experience.

For several years I had been subject to attacks of the singular disorder which physicians have agreed to term catalepsy, in default of a more definite title. Although both the immediate and the predisposing causes, and even the actual diagnosis, of this disease are still mysterious, its obvious and apparent character is sufficiently well understood. Its variations seem to be chiefly of degree. Sometimes the patient lies, for a day only, or even for a shorter period, in a species of exaggerated lethargy. He is senseless and externally motionless; but the pulsation of the heart is still faintly perceptible; some traces of warmth remain; a slight color lingers within the centre of the cheek; and, upon application of a mirror to the lips, we can detect a torpid, unequal, and vacillating action of the lungs. Then again the duration of the trance is for weeks—even for months; while the closest scrutiny, and the most rigorous medical tests, fail to establish any material distinction between the state of the sufferer and what we conceive of absolute death. Very usually, he is saved from premature interment solely by the knowledge of his friends that he has been previously subject to catalepsy, by the consequent suspicion excited, and, above all, by the non-appearance of decay. The advances of the malady are, luckily, gradual. The first manifestations, although marked, are unequivocal. The fits grow successively more and more distinctive, and endure each for a longer term than the preceding. In this lies the principal security from inhumation. The unfortunate whose *first* attack should be of the extreme character which is occasionally seen, would almost inevitably be consigned alive to the tomb.

My own case differed in no important particular from those mentioned in medical books. Sometimes, without any apparent cause, I sank, little by little, into a condition of hemi-syncope, or half swoon; and, in this condition, without pain, without ability to stir, or, strictly speaking, to think, but with a dull lethargic consciousness of life and of the presence of those who surrounded my bed, I remained, until the crisis of the disease restored me, suddenly, to perfect sensation. At other times I was quickly and impetuously smitten. I grew sick, and numb, and chilly, and dizzy, and so fell prostrate at once. Then, for weeks, all was void, and black, and silent, and Nothing became the universe. Total annihilation could be no more. From these latter attacks I awoke, however, with a gradation slow in proportion to the suddenness of the seizure. Just as the day dawns to the friendless and houseless beggar who roams the streets throughout the long desolate winter night—just so tardily—just so wearily—just so cheerily came back the light of the Soul to me.

Apart from the tendency to trance, however, my general health appeared to be good; nor could I perceive that it was at all affected by the one prevalent malady—unless, indeed, an idiosyncrasy in my ordinary *sleep* may be looked upon as superinduced. Upon awaking from slumber, I could never gain, at once, thorough possession of my senses, and always remained, for many minutes, in much bewilderment and perplexity;—the mental faculties in general, but the memory in especial, being in a condition of absolute abeyance.

In all that I endured there was no physical suffering, but of moral distress an infinitude. My fancy grew charnel. I talked "of worms, of tombs, and epitaphs." I was lost in reveries of death, and the idea of premature burial held continual possession of my brain. The ghastly Danger to which I was subjected haunted me day and night. In the former, the torture of meditation was excessive—in the latter, supreme. When the grim Darkness overspread the Earth, then, with every horror of thought, I shook—shook as the quivering plumes upon the hearse. When Nature could endure wakefulness no longer, it was with a struggle that I consented to sleep—for I shuddered to reflect that, upon awaking, I might find myself the tenant of a grave. And when, finally, I sank into slumber, it was only to rush at once into a world of phantasms, above which, with vast, sable, overshadowing wings, hovered, predominant, the one sepulchral Idea.

From the innumerable images of gloom which thus oppressed me in dreams, I select for record but a solitary vision. Methought I was immersed in a cataleptic trance of more than usual duration and profundity. Suddenly there came an icy hand upon my forehead, and an impatient, gibbering voice whispered the word "Arise!" within my ear.

I sat erect. The darkness was total. I could not see the figure of him who had aroused me. I could call to mind neither the period at which I had fallen into the trance, nor the locality in which I then lay. While I remained motionless, and busied in endeavors to collect my thoughts, the cold hand grasped me fiercely by the wrist, shaking it petulantly, while the gibbering voice said again:

"Arise! did I not bid thee arise?"

"And who," I demanded, "art thou?"

"I have no name in the regions which I inhabit," replied the voice mournfully; "I was mortal, but am fiend. I was merciless, but am pitiful. Thou dost feel that I shudder.—My teeth chatter as I speak, yet it is not with the chilliness of the night—of the night without end. But this hideousness is insufferable. How canst *thou* tranquilly sleep? I cannot rest for the cry of these great agonies. These sights are more than I can bear. Get thee up! Come with me into the outer Night, and let me unfold to thee the graves. Is not this a spectacle of woe?—Behold!"

I looked; and the unseen figure, which still grasped me by the wrist, had caused to be thrown open the graves of all mankind; and from each issued the faint phosphoric radiance of decay; so that I could see into the innermost recesses, and there view the shrouded bodies in their sad and solemn slumbers with the worm. But, alas! the real sleepers were fewer, by many millions, than those who slumbered not at all; and there was a feeble struggling; and there was a general and sad unrest; and from out the depths of the countless pits there came a melancholy rustling from the garments of the buried. And of those who seemed tranquilly to repose, I saw that a vast number had changed, in a greater or less degree, the rigid and uneasy position in which they had originally been entombed. And the voice again said to me, as I gazed:

"Is it not—oh! is it *not* a pitiful sight?"—but, before I could find words to reply, the figure had ceased to grasp my wrist, the phosphoric lights expired, and the graves were closed with a sudden violence, while from out them arose a tumult of despairing cries, saying again—"Is it not—oh, God! is it *not* a very pitiful sight?"

Phantasies such as these, presenting themselves at night, extended their terrific influence far into my waking hours.—My nerves became thoroughly unstrung, and I fell a prey to perpetual horror. I hesitated to ride, or to walk, or to indulge in any exercise that would carry me from home. In fact, I no longer dared trust myself out of the immediate presence of those who were aware of my proneness to catalepsy, lest, falling into one of my usual fits, I should be buried before my real condition could be ascertained. I doubted the care, the fidelity of my dearest friends. I dreaded that, in some trance of more than customary duration, they might be prevailed upon to regard me as irrecoverable. I even went so far as to fear that, as I occasioned much trouble, they might be glad to consider any very protracted attack as sufficient excuse for getting rid of me altogether. It was in vain they endeavored to reassure me by the most solemn promises. I exacted the most sacred oaths, that under no circumstances they would bury me until decomposition had so materially advanced as to render farther preservation impossible. And, even then, my mortal terrors would listen to no reason—would accept no consolation. I entered into a series of elaborate precautions. Among other things, I had the family vault so remodelled as to admit of being readily opened from within. The slightest pressure upon a long lever that extended far into the tomb would cause the iron portals to fly back. There were arrangements also for the free admission of air and light, and convenient receptacles for food and water, within immediate reach of the coffin intended for my reception. This coffin was warmly and softly padded, and was provided with a lid, fashioned upon the principle of the vault-door, with the addition of springs so contrived that the feeblest movement of the body would be sufficient to set it at liberty. Besides all this, there was suspended from the roof of the tomb, a large bell, the rope of which, it was designed, should extend through a hole in the coffin, and so be fastened to one of the hands of the corpse. But, alas! what avails the vigilance against the Destiny of man? Not even these well contrived securities sufficed to save from the uttermost agonies of living inhumation, a wretch to these agonies foredoomed!

There arrived an epoch—as often before there had arrived—in which I found myself emerging from total unconsciousness into the first feeble and indefinite sense of existence.—Slowly—with a tortoise gradation—approached the faint gray dawn of the psychal day. A torpid uneasiness. An apathetic endurance of dull pain. No care—no hope—no effort. Then, after a long interval, a ringing in the ears; then, after a lapse still longer, a pricking or tingling sensation in the extremities; then a seemingly eternal period of pleasurable quiescence, during which the awakening feelings are struggling into thought; then a brief re-sinking into non-entity; then a sudden recovery. At length the slight quivering of an eyelid, and immediately thereupon, an electric shock of a terror, deadly and indefinite, which sends the blood in torrents from the temples to the heart. And now the first positive effort to think. And now the first endeavor to remember. And now a partial and evanescent success. And now the memory has so far regained its dominion, that, in some measure, I am cognizant of my state. I feel that I am not awaking from ordinary sleep. I recollect that I have been subject to catalepsy. And now, at last, as if by the rush of an ocean, my shuddering spirit is overwhelmed by the one grim Danger—by the one spectral and ever-prevalent Idea.

For some minutes after this fancy possessed me, I remained without motion. And why? I could not summon courage to move. I dared not make the effort which was to satisfy me of my fate—and yet there was something at my heart which whispered me *it was sure*. Despair—such as no other species of wretchedness ever calls into being—despair alone urged me, after long irresolution, to uplift the heavy lids of my eyes. I uplifted them. It was dark—all dark. I knew that the fit was over. I knew that the crisis of my disorder had long passed. I knew that I had now fully recovered the use of my visual faculties—and yet it was dark—all dark—the intense and utter raylessness of the Night that endureth for evermore.

I endeavored to shriek; and my lips and my parched tongue moved convulsively together in the attempt—but no voice issued from the cavernous lungs, which, oppressed as if by the weight of some incumbent mountain, gasped and palpitated, with the heart, at every elaborate and struggling inspiration.

The movement of the jaws, in this effort to cry aloud, showed me that they were bound up, as is usual with the dead. I felt, too, that I lay upon some hard substance; and by something similar my sides were, also, closely compressed. So far, I had not ventured to stir any of my limbs—but now I violently threw up my arms, which had been lying at length, with the wrists crossed. They struck a solid wooden substance, which extended above my person at an elevation of not more than six inches from my face. I could no longer doubt that I reposed within a coffin at last.

And now, amid all my infinite miseries, came sweetly the cherub Hope—for I thought of my precautions. I writhed, and made spasmodic exertions to force open the lid: it would not move. I felt my wrists for the bell-rope: it was not to be found. And now the Comforter fled for ever, and a still sterner Despair reigned triumphant; for I could not help perceiving the absence of the paddings which I had so carefully prepared—and then, too, there came suddenly to my nostrils the strong peculiar odor of moist earth. The conclusion was irresistible. I was *not* within the vault. I had fallen into a trance while absent from home—while among strangers—when, or how, I could not remember—and it was they who had buried me as a dog—nailed up in some common coffin—and thrust, deep, deep, and forever, into some ordinary and nameless *grave.*

As this awful conviction forced itself, thus, into the innermost chambers of my soul, I once again struggled to cry aloud. And in this second endeavor I succeeded. A long, wild, and continuous shriek, or yell, of agony, resounded through the realms of the subterrene Night.

"Hillo! hillo, there!" said a gruff voice, in reply.

"What the devil's the matter now?" said a second.

"Get out o' that!" said a third.

"What do you mean by yowling in that ere kind of style, like a catty-mount?" said a fourth; and hereupon I was seized and shaken without ceremony, for several minutes, by a junto of very rough-looking individuals. They did not arouse me from my slumber—for I was wide awake when I screamed—but they restored me to the full possession of my memory.

This adventure occurred near Richmond, in Virginia. Accompanied by a friend, I had proceeded, upon a gunning expedition, some miles down the banks of the James River. Night approached, and we were overtaken by a storm. The cabin of a small sloop lying at anchor in the stream, and laden with garden mould, afforded us the only available shelter. We made the best of it, and passed the night on board. I slept in one of the only two berths in the vessel—and the berths of a sloop of sixty or seventy tons need scarcely be described. That which I occupied had no bedding of any kind. Its extreme width was eighteen inches. The distance of its bottom from the deck overhead was precisely the same. I found it a matter of exceeding difficulty to squeeze myself in. Nevertheless, I slept soundly; and the whole of my vision—for it was no dream, and no nightmare—arose naturally from the circumstances of my position—from my ordinary bias of thought—and from the difficulty, to which I have alluded, of collecting my senses, and especially of regaining my memory, for a long time after awaking from slumber. The men who shook me were the crew of the sloop, and some laborers engaged to unload it. From the load itself came the earthy smell. The bandage about the jaws was a silk handkerchief in which I had bound up my head, in default of my customary nightcap.

The tortures endured, however, were indubitably quite equal, for the time, to those of actual sepulture. They were fearfully—they were inconceivably hideous; but out of Evil proceeded Good; for their very excess wrought in my spirit an inevitable revulsion. My soul acquired tone—acquired temper. I went abroad. I took vigorous exercise. I breathed the free air of Heaven. I thought upon other subjects than Death. I discarded my medical books. "Buchan" I burned. I read no "Night Thoughts"—no fustian about church-yards—no bugaboo tales—*such as this.* In short I became a new man, and lived a man's life. From that memorable night, I dismissed forever my charnel apprehensions, and with them vanished the cataleptic disorder, of which, perhaps, they had been less the consequence than the cause.

There are moments when, even to the sober eye of Reason, the world of our sad Humanity may assume the semblance of a Hell—but the imagination of man is no Carathis, to explore with impunity its every cavern. Alas! the grim legion of sepulchral terrors cannot be regarded as altogether fanciful—but, like the Demons in whose company Afrasiab made his voyage down the Oxus, they must sleep, or they will devour us—they must be suffered to slumber, or we perish.

DISCUSSION QUESTIONS

1. The narrator opens the tale by sharing an allegedly true series of premature burial case studies. He then hoaxes his readers as the allegedly nonfiction piece morphs into a fictional tale. Describe how the tale would have been different had Poe written it as straight fiction. In health care, is there ever a thin line between fact and fiction—or might we at least argue that some information we accept as fact may be more complex when given further consideration? Is the information recorded in a patient's chart, for example, all "factual"?

2. When an illness or injury renders a patient unconscious—or conscious but unable to communicate by normal means—what special considerations must providers take in their care? In particular, how do the normal modes of communication shift? Consider watching a film like *The Diving Bell and the Butterfly* (2007) and discussing the roles various health care providers play in adapting new forms of communication.

3. The narrator claims, "The boundaries which divide Life from Death are at best shadowy and vague. Who shall say where one ends, and where the other begins?" Has the medico-scientific community made notable leaps in identifying the line between life and death since the nineteenth century?

4. When death seems imminent, what ethical considerations are involved in end-of-life decision-making? Describe the various roles the members of an interprofessional team, including chaplains and patient advocates, can play in end-of-life care to ensure both patient and family are cared for with dignity.

5. The narrator's illness disappears once he stops reading "medical books" and "bugaboo tales." What significance might we assign to this change? To what extent are our own perceptions of health and illness influenced by various media, which may include both fiction and nonfiction? Are health care providers and biomedical scientists able to remain objective despite these influences?

6. How would you characterize the behaviors of the members of the medical community described in this tale? What motivates them, and how do their motivations influence the patient-provider relationship? Would you describe their methods as patient-centered? Explain your answer.

Mesmeric Revelation

August 1844, *Columbian Magazine*

One of the most popular books on mesmerism during Poe's time was Facts in Mesmerism *(1840), by Reverend Chauncey Hare Townshend, who also happened to be a poet. Poe reviewed the book positively and relied upon it when writing "Mesmeric Revelation" (Lind 1947).*[1] *The book contained a supplement composed of testimonials of mesmerism, a non-fictional genre Poe mimicked with this fictional tale. However, in "Mesmeric Revelation," this testimony is related by a mesmerist-doctor who is called to the deathbed of a longtime patient, Mr. Vankirk, with whom he has a conversation about the nature of God and the universe. During this conversation, Vankirk describes a conception of the spiritual world that reflected Poe's own "faith," which he described in letters to James Russell Lowell and Dr. Thomas H. Chivers in the month preceding publication of the piece.*[2]

Like the eponymous character in "The Facts in the Case of M. Valdemar," Vankirk is dying of tuberculosis. And, like Valdemar, he willingly submits to the mesmeric process, although for different reasons. Because of the partnership between the mesmerist and his patient and the empowerment the patient seemingly gets through mesmerism, the story is also similar to "The Tale of the Ragged Mountains," in which a strong rapport has been established between patient and provider. Of the three mesmerism stories, however, "Mesmeric Revelation" is the only one narrated by the mesmerist. In fact, of all of the tales collected in this anthology, with the exception of the interior tale of Hawthorne's "The Haunted Quack," this is the only one narrated by a "doctor." Because of this unique feature, it is worth analyzing the narrator's arguments and considering whether he is a mouthpiece for Poe's own ideas or, perhaps, a vehicle for satirizing mesmerism. Like the other mesmerism stories, "Mesmeric Revelation" also invites us to revisit how the term "doctor" was defined in the early to mid-nineteenth century—and to consider the means by which healers of all professions, including chaplains and other spiritual advisors, provide patients with care and comfort in their dying hours.

•

WHATEVER doubt may still envelop the *rationale* of mesmerism, its startling *facts* are now almost universally admitted. Of these latter, those who doubt, are your mere doubters by profession—an unprofitable and disreputable tribe. There can be no more absolute waste of time than the attempt to *prove,* at the present day, that man, by mere exercise of will, can so impress his fellow, as to cast him into an abnormal condition, in which the phenomena resemble very closely those of *death,* or at least resemble them more nearly than they do the phenomena of any other normal condition within our cognizance; that, while in this state, the person so impressed employs only with effort, and then feebly, the external organs of sense, yet perceives, with keenly refined perception, and through channels supposed unknown, matters beyond the scope of the physical organs; that, moreover, his intellectual faculties are wonderfully exalted and invigorated; that his sympathies with the person so impressing him are profound; and, finally, that his susceptibility to the impression increases with its frequency, while, in the same proportion, the peculiar phenomena elicited are more extended and more *pronounced.*

I say that these—which are the laws of mesmerism in its general features—it would be supererogation to demonstrate; nor shall I inflict upon my readers so needless a demonstration to-day. My purpose at present is a very different one indeed. I am impelled, even in the teeth of a world of prejudice, to detail without comment the very remarkable substance of a colloquy, occurring between a sleep-waker and myself.

I had been long in the habit of mesmerizing the person in question, (Mr. Vankirk,) and the usual acute susceptibility and exaltation of the mesmeric perception had supervened. For many months he had been laboring under confirmed phthisis, the more distressing effects of which had been relieved by my manipulations; and on the night of Wednesday, the fifteenth instant, I was summoned to his bedside.

The invalid was suffering with acute pain in the region of the heart, and breathed with great difficulty, having all the ordinary symptoms of asthma. In spasms such as these he had usually found relief from the application of mustard to the nervous centres, but to-night this had been attempted in vain.

As I entered his room he greeted me with a cheerful smile, and although evidently in much bodily pain, appeared to be, mentally, quite at ease.

"I sent for you to-night," he said, "not so much to administer to my bodily ailment, as to satisfy me concerning certain psychal impressions which, of late, have occasioned me much anxiety and surprise. I need not tell you how sceptical I have hitherto been on the topic of the soul's immortality. I cannot deny that there has always existed, as if in that very soul which I have been denying, a vague half-sentiment of its own existence. But this half-sentiment at no time amounted to conviction. With it my reason had nothing to do. All attempts at logical inquiry resulted, indeed, in leaving me more sceptical than before. I had been advised to study Cousin. I studied him in his own works as well as in those of his European and American echoes. The 'Charles Elwood' of Mr. Brownson, for example, was placed in my hands. I read it with profound attention. Throughout I found it logical, but the portions which were not *merely* logical were unhappily the initial arguments of the disbelieving hero of the book. In his summing up it seemed evident to me that the reasoner had not even succeeded in convincing himself. His end had plainly forgotten his beginning, like the government of Trinculo. In short, I was not long in perceiving that if man is to be intellectually convinced of his own immortality, he will never be so convinced by the mere abstractions which have been so long the fashion of the moralists of England, of France, and of Germany. Abstractions may amuse and exercise, but take no hold on the mind. Here upon earth, at least, philosophy, I am persuaded, will always in vain call upon us to look upon qualities as things. The will may assent—the soul—the intellect, never.

"I repeat, then, that I only half felt, and never intellectually believed. But latterly there has been a certain deepening of the feeling, until it has come so nearly to resemble the acquiescence of reason, that I find it difficult to distinguish between the two. I am enabled, too, plainly to trace this effect to the mesmeric influence. I cannot better explain my meaning than by the hypothesis that the mesmeric exaltation enables me to perceive a train of ratiocination which, in my abnormal existence, convinces, but which, in full accordance with the mesmeric phenomena, does not extend, except through its *effect,* into my normal condition. In sleep-waking, the reasoning and its conclusion—the cause and its effect—are present together. In my natural state, the cause vanishing, the effect only, and perhaps only partially, remains.

"These considerations have led me to think that some good results might ensue from a series of well-directed questions propounded to me while mesmerized. You have often observed the profound self-cognizance evinced by the sleep-waker—the extensive knowledge he displays upon all points relating to the mesmeric condition itself; and from this self-cognizance may be deduced hints for the proper conduct of a catechism."

I consented of course to make this experiment. A few passes threw Mr. Vankirk into the mesmeric sleep. His breathing became immediately more easy, and he seemed to suffer no physical uneasiness. The following conversation then ensued:—V. in the dialogue representing the patient, and P. myself.

P. Are you asleep?

V. Yes—no; I would rather sleep more soundly.

P. [*After a few more passes.*] Do you sleep now?

V. Yes.

P. How do you think your present illness will result?

V. [*After a long hesitation and speaking as if with effort.*] I must die.

P. Does the idea of death afflict you?

V. [*Very quickly.*] No—no!

P. Are you pleased with the prospect?

V. If I were awake I should like to die, but now it is no matter. The mesmeric condition is so near death as to content me.

P. I wish you would explain yourself, Mr. Vankirk.

V. I am willing to do so, but it requires more effort than I feel able to make. You do not question me properly.

P. What then shall I ask?

V. You must begin at the beginning.

P. The beginning! but where is the beginning?

V. You know that the beginning is GOD. [*This was said in a low, fluctuating tone, and with every sign of the most profound veneration.*]

P. What then is God?

V. [*Hesitating for many minutes.*] I cannot tell.

P. Is not God spirit?

V. While I was awake I knew what you meant by "spirit," but now it seems only a word—such for instance as truth, beauty—a quality, I mean.

P. Is not God immaterial?

V. There is no immateriality—it is a mere word. That which is not matter, is not at all—unless qualities are things.

P. Is God, then, material?

V. No. [*This reply startled me very much.*]

P. What then is he?

V. [*After a long pause, and mutteringly.*] I see—but it is a thing difficult to tell. [*Another long pause.*] He is not spirit, for he exists. Nor is he matter, *as you understand it.* But there are *gradations* of matter of which man knows nothing; the grosser impelling the finer, the finer pervading the grosser. The atmosphere, for example, impels the electric principle, while the electric principle permeates the atmosphere. These gradations of matter increase in rarity or fineness, until we arrive at a matter *unparticled*—without particles—indivisible—*one*; and here the law of impulsion and permeation is modified. The ultimate, or unparticled matter, not only permeates all things, but impels all things—and thus *is* all things within itself. This matter is God. What men attempt to embody in the word "thought," is this matter in motion.

P. The metaphysicians maintain that all action is reducible to motion and thinking, and that the latter is the origin of the former.

V. Yes; and I now see the confusion of idea. Motion is the action of *mind*— not of *thinking*. The unparticled matter, or God, in quiescence, is (as nearly as we can conceive it) what men call mind. And the power of self-movement (equivalent in effect to human volition) is, in the unparticled matter, the result of its unity and omniprevalence; *how,* I know not, and now clearly see that I shall never know. But the unparticled matter, set in motion by a law or quality existing within itself, is thinking.

P. Can you give me no more precise idea of what you term the unparticled matter?

V. The matters of which man is cognizant escape the senses in gradation. We have, for example, a metal, a piece of wood, a drop of water, the atmosphere, a gas, caloric, electricity, the luminiferous ether. Now, we call all these things matter, and embrace all matter in one general definition; but in spite of this, there can be no two ideas more essentially distinct than that which we attach to a metal, and that which we attach to the luminiferous ether. When we reach the latter, we feel an almost irresistible inclination to class it with spirit, or with nihility. The only consideration which restrains us is our conception of its atomic constitution; and here, even, we have to seek aid from our notion of an atom, as something possessing in infinite minuteness, solidity, palpability, weight. Destroy the idea of the atomic constitution and we should no longer be able to regard the ether as an entity, or at least as matter. For want of a better word we might term it spirit. Take, now, a step beyond the luminiferous ether—conceive a matter as much more rare than the ether, as this ether is more rare than the metal, and we arrive at once (in spite of all the school dogmas) at a unique mass—an unparticled matter. For although we may admit infinite littleness in the atoms themselves, the infinitude of littleness in the spaces between them is an absurdity. There will be a point—there will be a degree of rarity, at which, if the atoms are sufficiently numerous, the interspaces must vanish, and the mass absolutely coalesce. But the consideration of the atomic constitution being now taken away, the nature of the mass inevitably glides into what we conceive of spirit. It is clear, however, that it is as fully matter as before. The truth is, it is impossible to conceive spirit, since it is impossible to imagine what is not. When we flatter ourselves that we have formed its conception, we have merely deceived our understanding by the consideration of infinitely rarefied matter.

P. There seems to me an insurmountable objection to the idea of absolute coalescence;—and that is the very slight resistance experienced by the heavenly bodies in their revolutions through space—a resistance now ascertained, it is true, to exist in *some* degree, but which is, nevertheless, so slight as to have been quite overlooked by the sagacity even of Newton. We know that the resistance of bodies is, chiefly, in proportion to their density. Absolute coalescence is absolute density. Where there are no interspaces, there can be no yielding. An ether, absolutely dense, would put an infinitely more effectual stop to the progress of a star than would an ether of adamant or of iron.

V. Your objection is answered with an ease which is nearly in the ratio of its apparent unanswerability.—As regards the progress of the star, it can make no difference whether the star passes through the ether *or the ether through it.* There is no astronomical error more unaccountable than that which reconciles the known retardation of the comets with the idea of their passage through an ether: for, however rare this ether be supposed, it would put a stop to all sidereal revolution in a very far briefer period than has been admitted by those astronomers who have endeavored to slur over a point which they found it impossible to comprehend. The retardation actually experienced is, on the other hand, about that which might be expected from the *friction* of the ether in the instantaneous passage through the orb. In the one case, the retarding force is momentary and complete within itself—in the other it is endlessly accumulative.

P. But in all this—in this identification of mere matter with God—is there nothing of irreverence? [*I was forced to repeat this question before the sleep-waker fully comprehended my meaning.*]

V. Can you say *why* matter should be less reverenced than mind? But you forget that the matter of which I speak is, in all respects, the very "mind" or "spirit" of the schools, so far as regards its high capacities, and is, moreover, the "matter" of these schools at the same time. God, with all the powers attributed to spirit, is but the perfection of matter.

P. You assert, then, that the unparticled matter, in motion, is thought?

V. In general, this motion is the universal thought of the universal mind. This thought creates. All created things are but the thoughts of God.

P. You say, "in general."

V. Yes. The universal mind is God. For new individualities, *matter* is necessary.

P. But you now speak of "mind" and "matter" as do the metaphysicians.

V. Yes—to avoid confusion. When I say "mind," I mean the unparticled or ultimate matter; by "matter," I intend all else.

P. You were saying that "for new individualities matter is necessary."

V. Yes; for mind, existing unincorporate, is merely God. To create individual, thinking beings, it was necessary to incarnate portions of the divine mind. Thus man is individualized. Divested of corporate investiture, he were God. Now the particular motion of the incarnated portions of the unparticled matter is the thought of man; as the motion of the whole is that of God.

P. You say that divested of the body man will be God?

V. [*After much hesitation.*] I could not have said this; it is an absurdity.

P. [*Referring to my notes.*] You *did* say that "divested of corporate investiture man were God."

V. And this is true. Man thus divested *would be* God—would be unindividualized. But he can never be thus divested—at least never *will be*—else we must imagine an action of God returning upon itself—a purposeless and futile action. Man is a creature. Creatures are thoughts of God. It is the nature of thought to be irrevocable.

P. I do not comprehend. You say that man will never put off the body?

V. I say that he will never be bodiless.

P. Explain.

V. There are two bodies—the rudimental and the complete, corresponding with the two conditions of the worm and the butterfly. What we call "death," is but the painful metamorphosis. Our present incarnation is progressive, preparatory, temporary. Our future is perfected, ultimate, immortal. The ultimate life is the full design.

P. But of the worm's metamorphosis we are palpably cognizant.

V. We, certainly—but not the worm. The matter of which our rudimental body is composed, is within the ken of the organs of that body; or, more distinctly, our rudimental organs are adapted to the matter of which is formed the rudimental body; but not to that of which the ultimate is composed. The ultimate body thus escapes our rudimental senses, and we perceive only the shell which falls, in decaying, from the inner form; not that inner form itself; but this inner form, as well as the shell, is appreciable by those who have already acquired the ultimate life.

P. You have often said that the mesmeric state very nearly resembles death. How is this?

V. When I say that it resembles death, I mean that it resembles the ultimate life; for when I am entranced the senses of my rudimental life are in abeyance, and I perceive external things directly, without organs, through a medium which I shall employ in the ultimate, unorganized life.

P. Unorganized?

V. Yes; organs are contrivances by which the individual is brought into sensible relation with particular classes and forms of matter, to the exclusion of other classes and forms. The organs of man are adapted to his rudimental condition, and to that only; his ultimate condition, being unorganized, is of unlimited comprehension in all points but one—the nature of the volition of God—that is to say, the motion of the unparticled matter. You will have a distinct idea of the ultimate body by conceiving it to be entire brain. This it is *not*; but a conception of this nature will bring you near a comprehension of what it *is*. A luminous body imparts vibration to the luminiferous ether. The vibrations generate similar ones within the retina; these again communicate similar ones to the optic nerve. The nerve conveys similar ones to the brain; the brain, also, similar ones to the unparticled matter which permeates it. The motion of this latter is thought, of which perception is the first undulation. This is the mode by which the mind of the rudimental life communicates with the external world; and this external world is, to the rudimental life, limited, through the idiosyncrasy of its organs. But in the ultimate, unorganized life, the external world reaches the whole body, (which is of a substance having affinity to brain, as I have said,) with no other intervention than that of an infinitely rarer ether than even the luminiferous; and to this ether—in unison with it—the whole body vibrates, setting in motion the unparticled matter which permeates it. It is to the absence of idiosyncratic organs, therefore, that we must attribute the nearly unlimited perception of the ultimate life. To rudimental beings, organs are the cages necessary to confine them until fledged.

P. You speak of rudimental "beings." Are there other rudimental thinking beings than man?

V. The multitudinous conglomeration of rare matter into nebulæ, planets, suns, and other bodies which are neither nebulæ, suns, nor planets, is for the sole purpose of supplying *pabulum* for the idiosyncrasy of the organs of an infinity of rudimental beings. But for the necessity of the rudimental, prior to the ultimate life, there would have been no bodies such as these. Each of these is tenanted by a distinct variety of organic, rudimental, thinking creatures. In all, the organs vary with the features of the place tenanted. At death, or metamorphosis, these creatures, enjoying the ultimate life—immortality—and cognizant of all secrets but *the one,* act all things and pass everywhere by mere volition:—indwelling, not the stars, which to us seem the sole palpabilities, and for the accommodation of which we blindly deem space created—but that SPACE itself—that infinity of which the truly substantive vastness swallows up the star-shadows—blotting them out as non-entities from the perception of the angels.

P. You say that "but for the *necessity* of the rudimental life, there would have been no stars." But why this necessity?

V. In the inorganic life, as well as in the inorganic matter generally, there is nothing to impede the action of one simple *unique* law—the Divine Volition.

With the view of producing impediment, the organic life and matter (complex, substantial, and law-encumbered,) were contrived.

P. But again—why need this impediment have been produced?

V. The result of law inviolate is perfection—right—negative happiness. The result of law violate is imperfection, wrong, positive pain. Through the impediments afforded by the number, complexity, and substantiality of the laws of organic life and matter, the violation of law is rendered, to a certain extent, practicable. Thus pain, which in the inorganic life is impossible, is possible in the organic.

P. But to what good end is pain thus rendered possible?

V. All things are either good or bad by comparison. A sufficient analysis will show that pleasure, in all cases, is but the contrast of pain. *Positive* pleasure is a mere idea. To be happy at any one point we must have suffered at the same. Never to suffer would have been never to have been blessed. But it has been shown that, in the inorganic life, pain cannot be; thus the necessity for the organic. The pain of the primitive life of Earth, is the sole basis of the bliss of the ultimate life in Heaven.

P. Still, there is one of your expressions which I find it impossible to comprehend—"the truly *substantive* vastness of infinity."

V. This, probably, is because you have no sufficiently generic conception of the term *"substance"* itself. We must not regard it as a quality, but as a sentiment:—it is the perception, in thinking beings, of the adaptation of matter to their organization. There are many things on the Earth, which would be nihility to the inhabitants of Venus—many things visible and tangible in Venus, which we could not be brought to appreciate as existing at all. But to the inorganic beings—to the angels—the whole of the unparticled matter is substance; that is to say, the whole of what we term "space" is to them the truest substantiality;—the stars, meantime, through what we consider their materiality, escaping the angelic sense, just in proportion as the unparticled matter, through what we consider its immateriality, eludes the organic.

As the sleep-waker pronounced these latter words, in a feeble tone, I observed on his countenance a singular expression, which somewhat alarmed me, and induced me to awake him at once. No sooner had I done this, than, with a bright smile irradiating all his features, he fell back upon his pillow and expired. I noticed that in less than a minute afterward his corpse had all the stern rigidity of stone. His brow was of the coldness of ice. Thus, ordinarily, should it have appeared, only after long pressure from Azrael's hand. Had the sleep-waker, indeed, during the latter portion of his discourse, been addressing me from out the region of the shadows?

DISCUSSION QUESTIONS

1. Describe the relationship between Dr. Vankirk and his patient. Who has more power in the relationship, if either of them? Do they have what you would consider a positive provider-patient relationship? Would you describe the care as patient- or relationship-centered? Does the relationship they've developed lead to better patient care and satisfaction?

2. This tale opens with a brief defense of mesmerism, in which the narrator accuses any remaining doubters of the reform as being "disreputable." What effect does this rhetorical claim have on you as a reader? What effect do you think it had on a nineteenth-century American reader? What types of rhetoric are used in health care today to either convince or dissuade the public about the validity of certain health care practices? Conduct a rhetorical analysis of a piece of health care literature—an academic article, an informational brochure, a history-taking checklist—asking yourself what the purpose of the document is, what methods of persuasion are used, and what the piece of writing says about the culture that produced it.

3. The patient, Mr. Vankirk, tells the mesmerist that he has called him not for relief of physical ailments but for a type of spiritual healing that he will gain through the mesmeric trance. What is it that the patient gains while in the trance that offers him comfort? What might this desire for spiritual comfort suggest about the needs of those close to death? Do you consider those who provide this comfort part of the interprofessional health care team? Research the spirituality and health movement in health care today. Outline the main philosophies that form the foundation of the movement; how do they relate to or depart from the spiritual content of this tale?

4. Write a paper comparing the mesmerist's treatment of his patient to Aylmer's treatment of Georgiana in Hawthorne's "The Birthmark" or Rappaccini's treatment of his daughter in "Rappaccini's Daughter."

5. Locate a contemporary poem or song dealing with end-of-life issues. How does the piece reflect twenty-first century attitudes toward death, dying, and end-of-life care? How do these attitudes compare to those of the nineteenth century? How have changes in the health care environment affected end-of-life health care delivery? Give particular consideration to issues related to patient authority and autonomy as well as interprofessionalism.

Some Words with a Mummy

April 1845, *American Whig Review*

Poe was inspired by the Egyptomania that swept across the Western world in the nineteenth century after Napoleon's 1789 invasion of Egypt, which led to the discovery of the Rosetta Stone. As scholars rushed to investigate Egyptian culture, numerous artifacts, including mummies, made their way to European and American museums (Montgomery 2012) as well as into lecture halls and, some claim, private homes, as mummy-unwrapping parties came into vogue (Fletcher 2011). According to Mabbott, the young Poe may have seen a mummy in a Virginia museum as early as 1823 (Poe 1978).[1]

As an adult, Poe pursued Egyptian themes in writing. An admirer of Champollion, the man who translated the Rosetta Stone, Poe demonstrated his own surpassing skill in code-cracking and incorporated codes and cryptograms into numerous works (Montgomery 2012).[2] In "Some Words with a Mummy," which may have been inspired by a tale titled "Letter from a Revived Mummy" that appeared in the New York Mirror *in January 1832 (King 1930, as cited in Forclaz 1968), he provides insights into Egyptian and American culture, humorously satirizing American's overblown sense of progress. He does so by telling the story of Doctor Ponnonner and his friends, who gather to unwrap and study a mummy the doctor has acquired. The action that ensues involves dissection, galvanic shock, and an extended conversation about embalming, making the tale a suitable fit for this collection. Because the doctor, possibly a version of Poe's New York physician Dr. John W. Francis (Sloane 1966) is the instigator of the experiments with the mummy, the tale raises questions about the perceptions of the role of the doctor in nineteenth-century society as well as issues related to power and professionalism within the evolving medical field and the social hierarchy. It also prompts us to reflect on the importance of cultural awareness and sensitivity in health care. Finally, like a number of Poe's works, it calls us to question the ethics of various ways the deceased have been—and are—used to further medical knowledge.*

•

THE symposium of the preceding evening had been a little too much for my nerves. I had a wretched head-ache, and was desperately drowsy. Instead of going out, therefore, to spend the evening, as I had proposed, it occurred to me that I could not do a wiser thing than just eat a mouthful of supper and go immediately to bed.

A *light* supper, of course. I am exceedingly fond of Welsh rabbit. More than a pound at once, however, may not at all times be advisable. Still, there can be no material objection to two. And really between two and three, there is merely a single unit of difference. I ventured, perhaps, upon four. My wife will have it five;—but, clearly, she has confounded two very distinct affairs. The abstract number, five, I am willing to admit; but, concretely, it has reference to bottles of Brown Stout, without which, in the way of condiment, Welsh rabbit is to be eschewed.

Having thus concluded a frugal meal, and donned my night-cap, with the serene hope of enjoying it till noon the next day, I placed my head upon the pillow, and, through the aid of a capital conscience, fell into a profound slumber forthwith.

But when were the hopes of humanity fulfilled? I could not have completed my third snore when there came a furious ringing at the street-door bell, and then an impatient thumping at the knocker, which awakened me at once. In a minute afterward and while I was still rubbing my eyes, my wife thrust in my face a note from my old friend, Doctor Ponnonner. It ran thus:

> Come to me, by all means, my dear good friend, as soon as you receive this. Come and help us to rejoice. At last, by long persevering diplomacy, I have gained the assent of the Directors of the City Museum, to my examination of the Mummy—you know the one I mean. I have permission to unswathe it and open it, if desirable. A few friends only will be present—you, of course. The Mummy is now at my house, and we shall begin to unroll it at eleven to-night.
>
> Yours ever,
>
> PONNONNER

By the time I had reached the "Ponnonner," it struck me that I was as wide awake as a man need be. I leaped out of bed in an ecstasy, overthrowing all in my way; dressed myself with a rapidity truly marvellous; and set off, at the top of my speed, for the Doctor's.

There I found a very eager company assembled. They had been awaiting me with much impatience; the Mummy was extended upon the dining table; and the moment I entered its examinations were commenced.

It was one of a pair brought, several years previously, by Captain Arthur Sabretash, a cousin of Ponnonner's, from a tomb near Eleithias, in the Libyan mountains, a considerable distance above Thebes on the Nile. The grottoes at this point, although less magnificent than the Theban sepulchres, are of higher interest, on account of affording more numerous illustrations of the private life of the Egyptians. The chamber from which our specimen was taken, was said to be very rich in such illustrations; the walls being completely covered with fresco paintings and bas-reliefs, while statues, vases, and Mosaic work of rich patterns, indicated the vast wealth of the deceased.

The treasure had been deposited in the Museum precisely in the same condition in which Captain Sabretash had found it;—that is to say, the coffin had not been disturbed. For eight years it had thus stood, subject only externally to public inspection. We had now, therefore, the complete Mummy at our disposal; and to those who are aware how very rarely the unransacked antique reaches our shores, it will be evident, at once, that we had great reason to congratulate ourselves upon our good fortune.

Approaching the table, I saw on it a large box, or case, nearly seven feet long, and perhaps three feet wide, by two feet and a half deep. It was oblong—not coffin-shaped. The material was at first supposed to be the wood of the sycamore (*platanus*), but, upon cutting into it, we found it to be pasteboard, or, more properly, *papier mâché*, composed of papyrus. It was thickly ornamented with paintings, representing funeral scenes, and other mournful subjects—interspersed among which, in every variety of position, were certain series of hieroglyphical characters, intended, no doubt, for the name of the departed. By good luck, Mr. Gliddon formed one of our party; and he had no difficulty in translating the letters, which were simply phonetic, and represented the word, *Allamistakeo*.

We had some difficulty in getting this case open without injury, but, having at length accomplished the task, we came to a second, coffin-shaped, and very considerably less in size than the exterior one, but resembling it precisely in every other respect. The interval between the two was filled with resin, which had, in some degree, defaced the colors of the interior box.

Upon opening this latter (which we did quite easily,) we arrived at a third case, also coffin-shaped, and varying from the second one in no particular, except in that of its material, which was cedar, and still emitted the peculiar and highly aromatic odor of that wood. Between the second and the third case there was no interval; the one fitting accurately within the other.

Removing the third case, we discovered and took out the body itself. We had expected to find it, as usual, enveloped in frequent rolls, or bandages, of linen, but, in place of these, we found a sort of sheath, made of papyrus, and coated with a layer of plaster, thickly gilt and painted. The paintings represented subjects connected with the various supposed duties of the soul, and its presentation to different divinities, with numerous identical human figures, intended, very probably, as portraits of the persons embalmed. Extending from head to foot was a columnar, or perpendicular, inscription, in phonetic hieroglyphics, giving again his name and titles, and the names and titles of his relations.

Around the neck thus unsheathed, was a collar of cylindrical glass beads, diverse in color, and so arranged as to form images of deities, of the scarabæus, etc., with the winged globe. Around the small of the waist was a similar collar, or belt.

Stripping off the papyrus, we found the flesh in excellent preservation, with no perceptible odor. The color was reddish. The skin was hard, smooth, and glossy. The teeth and hair were in good condition. The eyes (it seemed) had been removed, and glass ones substituted, which were very beautiful and wonderfully life-like, with the exception of somewhat too determined a stare. The finger and the toe nails were brilliantly gilded.

Mr. Gliddon was of opinion, from the redness of the epidermis, that, the embalmment had been effected altogether by asphaltum; but, on scraping the surface with a steel instrument, and throwing into the fire some of the powder thus obtained, the flavor of camphor and other sweet-scented gums became apparent.

We searched the corpse very carefully for the usual openings through which the entrails are extracted, but, to our surprise, we could discover none. No member of the party was at that period aware that entire or unopened mummies are not unfrequently met. The brain it was customary to withdraw through the nose; the intestines through an incision in the side; the body was then shaved, washed, and salted; then laid aside for several weeks, when the operation of embalming, properly so called, began.

As no trace of an opening could be found, Doctor Ponnonner was preparing his instruments for dissection, when I observed that it was then past two o'clock. Hereupon it was agreed to postpone the internal examination until the next evening; and we were about to separate for the present, when some one suggested an experiment or two with the Voltaic pile.

The application of electricity to a Mummy three or four thousand years old at the least, was an idea, if not very sage, still sufficiently original, and we all caught at it at once. About one tenth in earnest and nine tenths in jest, we arranged a battery in the Doctor's study, and conveyed thither the Egyptian.

It was only after much trouble that we succeeded in laying bare some portions of the temporal muscle which appeared of less stony rigidity than other parts of the frame, but which, as we had anticipated, of course, gave no indication of galvanic susceptibility when brought in contact with the wire. This, the first trial, indeed, seemed decisive, and, with a hearty laugh at our own absurdity, we were bidding each other good night, when my eyes, happening to fall upon those of the Mummy, were there immediately riveted in amazement. My brief glance, in fact, had sufficed to assure me that the orbs which we had all supposed to be glass, and which were originally noticeable for a certain wild stare, were now so far covered by the lids, that only a small portion of the *tunica albuginea* remained visible.

With a shout I called attention to the fact, and it became immediately obvious to all.

I cannot say that I was *alarmed* at the phenomenon, because "alarmed" is, in my case, not exactly the word. It is possible, however, that, but for the Brown Stout, I might have been a little nervous. As for the rest of the company, they really made no attempt at concealing the downright fright which possessed them. Doctor Ponnonner was a man to be pitied. Mr. Gliddon, by some peculiar process, rendered himself invisible. Mr. Silk Buckingham, I fancy, will scarcely be so bold as to deny that he made his way, upon all fours, under the table.

After the first shock of astonishment, however, we resolved, as a matter of course, upon further experiment forthwith. Our operations were now directed against the great toe of the right foot. We made an incision over the outside of the exterior *os sesamoideum pollicis pedis,* and thus got at the root of the *abductor* muscle. Re-adjusting the battery, we now applied the fluid to the bisected nerves—when, with a movement of exceeding life-likeness, the Mummy first drew up its right knee so as to bring it nearly in contact with the abdomen, and then, straightening the limb with inconceivable force, bestowed a kick upon Doctor Ponnonner, which had the effect of discharging that gentleman, like an arrow from a catapult, through a window into the street below.

We rushed out *en masse* to bring in the mangled remains of the victim, but had the happiness to meet him upon the staircase, coming up in an unaccountable hurry, brimfull of the most ardent philosophy, and more than ever impressed with the necessity of prosecuting our experiment with rigor and with zeal.

It was by his advice, accordingly, that we made, upon the spot, a profound incision into the tip of the subject's nose, while the Doctor himself, laying violent hands upon it, pulled it into vehement contact with the wire.

Morally and physically—figuratively and literally—was the effect electric. In the first place, the corpse opened its eyes and winked very rapidly for several minutes as does Mr. Barnes in the pantomime; in the second place, it sneezed; in the third, it sat upon end; in the fourth, it shook its fist in Doctor Ponnonner's face; in the fifth, turning to Messieurs Gliddon and Buckingham, it addressed them, in very capital Egyptian, thus:

"I must say, gentlemen, that I am as much surprised as I am mortified at your behavior. Of Doctor Ponnonner nothing better was to be expected. He is a poor little fat fool who *knows* no better. I pity and forgive him. But you, Mr. Gliddon—and you, Silk—who have travelled and resided in Egypt until one might imagine you to the manor born—you, I say, who have been so much among us that you speak Egyptian fully as well, I think, as you write your mother tongue—you, whom I have always been led to regard as the firm friend of the mummies—I really did anticipate more gentlemanly conduct from *you*. What am I to think of your standing quietly by and seeing me thus unhandsomely used? What am I to suppose by your permitting Tom, Dick, and Harry to strip me of my coffins, and my clothes, in this wretchedly cold climate? In what light (to come to the point) am I to regard your aiding and abetting that miserable little villain, Doctor Ponnonner, in pulling me by the nose?"

It will be taken for granted, no doubt, that upon hearing this speech under the circumstances, we all either made for the door, or fell into violent hysterics, or went off in a general swoon. One of these three things was, I say, to be expected. Indeed each and all of these lines of conduct might have been very plausibly pursued. And, upon my word, I am at a loss to know how or why it was that we pursued neither the one nor the other. But, perhaps, the true reason is to be sought in the spirit of the age, which proceeds by the rule of contraries altogether, and is now usually admitted as the solution of everything in the way of paradox and impossibility. Or, perhaps, after all, it was only the Mummy's exceedingly natural and matter-of-course air that divested his words of the terrible. However this may be, the facts are clear, and no member of our party betrayed any very particular trepidation, or seemed to consider that any thing had gone very especially wrong.

For my part I was convinced it was all right, and merely stepped aside, out of the range of the Egyptian's fist. Doctor Ponnonner thrust his hands into his breeches' pockets, looked hard at the Mummy, and grew excessively red in the face. Mr. Gliddon stroked his whiskers and drew up the collar of his shirt. Mr. Buckingham hung down his head, and put his right thumb into the left corner of his mouth.

The Egyptian regarded him with a severe countenance for some minutes, and at length, with a sneer, said:

"Why don't you speak, Mr. Buckingham? Did you hear what I asked you, or not? *Do* take your thumb out of your mouth!"

Mr. Buckingham, hereupon, gave a slight start, took his right thumb out of the left corner of his mouth, and, by way of indemnification, inserted his left thumb in the right corner of the aperture above-mentioned.

Not being able to get an answer from Mr. B., the figure turned peevishly to Mr. Gliddon, and, in a peremptory tone, demanded in general terms what we all meant.

Mr. Gliddon replied at great length, in phonetics; and but for the deficiency of American printing-offices in hieroglyphical type, it would afford me much pleasure to record here, in the original, the whole of his very excellent speech.

I may as well take this occasion to remark, that all the subsequent conversation in which the Mummy took a part, was carried on in primitive Egyptian, through the medium (so far as concerned myself and other untravelled members of the company)—through the medium, I say, of Messieurs Gliddon and Buckingham, as interpreters. These gentlemen spoke the mother-tongue of the mummy with inimitable fluency and grace; but I could not help observing that (owing, no doubt, to the introduction of images entirely modern, and, of course, entirely novel to the stranger) the two travellers were reduced, occasionally, to the employment of sensible forms for the purpose of conveying a particular meaning. Mr. Gliddon, at one period, for example, could not make the Egyptian comprehend the term "politics," until he sketched upon the wall, with a bit of charcoal, a little carbuncle nosed gentleman, out at elbows, standing upon a stump, with his left leg drawn back, his right arm thrown forward, with his fist shut, the eyes rolled up toward Heaven, and the mouth open at an angle of ninety degrees. Just in the same way Mr. Buckingham failed to convey the absolutely modern idea, "wig," until (at Doctor Ponnonner's suggestion) he grew very pale in the face, and consented to take off his own.

It will be readily understood that Mr. Gliddon's discourse turned chiefly upon the vast benefits accruing to science from the unrolling and disembowelling of mummies; apologizing, upon this score, for any disturbance that might have been occasioned *him,* in particular, the individual Mummy called Allamistakeo; and concluding with a mere hint (for it could scarcely be considered more,) that, as these little matters were now explained, it might be as well to proceed with the investigation intended. Here Doctor Ponnonner made ready his instruments.

In regard to the latter suggestions of the orator, it appears that Allamistakeo had certain scruples of conscience, the nature of which I did not distinctly learn; but he expressed himself satisfied with the apologies tendered, and, getting down from the table, shook hands with the company all round.

When this ceremony was at an end, we immediately busied ourselves in repairing the damages which our subject had sustained from the scalpel. We sewed up the wound in his temple, bandaged his foot, and applied a square inch of black plaster to the tip of his nose.

It was now observed that the Count, (this was the title, it seems, of Allamistakeo) had a slight fit of shivering—no doubt from the cold. The doctor immediately repaired to his wardrobe, and soon returned with a black dress coat, made in Jennings' best manner, a pair of sky-blue plaid pantaloons with straps, a pink gingham *chemise,* a flapped vest of brocade, a white sack overcoat, a walking cane with a hook, a hat with no brim, patent-leather boots, straw-colored kid gloves, an eye-glass, a pair of whiskers, and a waterfall cravat. Owing to the disparity of size between the Count and the doctor (the proportion being as two to one,) there was some little difficulty in adjusting these habiliments upon the person of the Egyptian; but when all was arranged, he might have been said to be dressed. Mr. Gliddon, therefore, gave him his arm, and led him to a comfortable chair by the fire, while the doctor rang the bell upon the spot and ordered a supply of cigars and wine.

The conversation soon grew animated. Much curiosity was, of course, expressed in regard to the somewhat remarkable fact of Allamistakeo's still remaining alive.

"I should have thought," observed Mr. Buckingham, "that it is high time you were dead."

"Why," replied the Count, very much astonished, "I am little more than seven hundred years old! My father lived a thousand, and was by no means in his dotage when he died."

Here ensued a brisk series of questions and computations, by means of which it became evident that the antiquity of the Mummy had been grossly misjudged. It had been five thousand and fifty years and some months since he had been consigned to the catacombs at Eleithias.

"But my remark," resumed Mr. Buckingham, "had no reference to your age at the period of interment; (I am willing to grant, in fact, that you are still a young man,) and my allusion was to the immensity of time during which, by your own showing, you must have been done up in asphaltum."

"In what?" said the Count.

"In asphaltum," persisted Mr. B.

"Ah, yes; I have some faint notion of what you mean; it might be made to answer, no doubt,—but in my time we employed scarcely any thing else than the Bichloride of Mercury."

"But what we are especially at a loss to understand," said Doctor Ponnonner, "is how it happens that, having been dead and buried in Egypt five thousand years ago, you are here to-day all alive, and looking so delightfully well."

"Had I been, as you say, *dead*," replied the Count, "it is more than probable that dead I should still be; for I perceive you are yet in the infancy of Galvanism, and cannot accomplish with it what was a common thing among us in the old days. But the fact is, I fell into catalepsy, and it was considered by my best friends that I was either dead or should be; they accordingly embalmed me at once—I presume you are aware of the chief principle of the embalming process?"

"Why, not altogether."

"Ah, I perceive;—a deplorable condition of ignorance! Well, I cannot enter into details just now: but it is necessary to explain that to embalm (properly speaking,) in Egypt, was to arrest indefinitely *all* the animal functions subjected to the process. I use the word 'animal' in its widest sense, as including the physical not more than the moral and *vital* being. I repeat that the leading principle of embalmment consisted, with us, in the immediately arresting, and holding in perpetual *abeyance,* all the animal functions subjected to the process. To be brief, in whatever condition the individual was, at the period of embalmment, in that condition he remained. Now, as it is my good fortune to be of the blood of the Scarabæus, I was embalmed *alive,* as you see me at present."

"The blood of the Scarabæus!" exclaimed Doctor Ponnonner.

"Yes. The Scarabæus was the *insignium,* or the 'arms,' of a very distinguished and very rare patrician family. To be 'of the blood of the Scarabæus,' is merely to be one of that family of which the Scarabæus is the *insignium.* I speak figuratively."

"But what has this to do with your being alive?"

"Why, it is the general custom, in Egypt, to deprive a corpse, before embalmment, of its bowels and brains; the race of the Scarabæi alone did not coincide with the custom. Had I not been a Scarabæus, therefore, I should have been without bowels and brains; and without either it is inconvenient to live."

"I perceive that," said Mr. Buckingham, "and I presume that all the *entire* mummies that come to hand are of the race of Scarabæi."

"Beyond doubt."

"I thought," said Mr. Gliddon very meekly, "that the Scarabæus was one of the Egyptian gods."

"One of the Egyptian *what?*" exclaimed the Mummy, starting to its feet.

"Gods!" repeated the traveler.

"Mr. Gliddon, I really am astonished to hear you talk in this style," said the Count, resuming his chair. "No nation upon the face of the earth has ever acknowledged more than *one god*. The Scarabæus, the Ibis, etc., were with us (as similar creatures have been with others) the symbols, or *media*, through which we offered worship to the Creator too august to be more directly approached."

There was here a pause. At length the colloquy was renewed by Doctor Ponnonner.

"It is not improbable, then, from what you have explained," said he, "that among the catacombs near the Nile there may exist other mummies of the Scarabæus tribe, in a condition of vitality."

"There can be no question of it," replied the Count; "all the Scarabæi embalmed accidentally while alive, are alive now. Even some of those *purposely* so embalmed, may have been overlooked by their executors, and still remain in the tombs."

"Will you be kind enough to explain," I said, "what you mean by 'purposely so embalmed?'"

"With great pleasure," answered the Mummy, after surveying me leisurely through his eye-glass—for it was the first time I had ventured to address him a direct question.

"With great pleasure," said he. "The usual duration of man's life, in my time, was about eight hundred years. Few men died, unless by most extraordinary accident, before the age of six hundred; few lived longer than a decade of centuries; but eight were considered the natural term. After the discovery of the embalming principle, as I have already described it to you, it occurred to our philosophers that a laudable curiosity might be gratified, and, at the same time, the interests of science much advanced, by living this natural term in instalments. In the case of history, indeed, experience demonstrated that something of this kind was indispensable. An historian, for example, having attained the age of five hundred, would write a book with great labor and then get himself carefully embalmed; leaving instructions to his executors *pro tem.*, that they should cause him to be revivified after the lapse of a certain period—say five or six hundred years. Resuming existence at the expiration of this time, he would invariably find his great work converted into a species of hap-hazard note-book—that is to say, into a kind of literary arena for the conflicting guesses, riddles, and personal squabbles of whole herds of exasperated commentators. These guesses, etc., which passed under the name of annotations, or emendations, were found so completely to have enveloped, distorted, and overwhelmed the text, that the author had to go about with a lantern to discover his own book. When discovered, it was never worth the trouble of the search. After rewriting it throughout, it was regarded as the bounden duty of the historian to set himself to work immediately in correcting, from his own private knowledge and experience, the traditions of the day concerning the epoch at which he had originally lived. Now this process of re-scription and personal rectification, pursued by various individual sages, from time to time, had the effect of preventing our history from degenerating into absolute fable."

"I beg your pardon," said Doctor Ponnonner at this point, laying his hand gently upon the arm of the Egyptian—"I beg your pardon, sir, but may I presume to interrupt you for one moment?"

"By all means, *sir,*" replied the Count, drawing up.

"I merely wished to ask you a question," said the Doctor. "You mentioned the historian's personal correction of *traditions* respecting his own epoch. Pray, sir, upon an average, what proportion of these Kabbala were usually found to be right?"

"The Kabbala, as you properly term them, sir, were generally discovered to be precisely on a par with the facts recorded in the un-re-written histories themselves;—that is to say, not one individual iota of either was ever known, under any circumstances, to be not totally and radically wrong."

"But since it is quite clear," resumed the Doctor, "that at least five thousand years have elapsed since your entombment, I take it for granted that your histories at that period, if not your traditions, were sufficiently explicit on that one topic of universal interest, the Creation, which took place, as I presume you are aware, only about ten centuries before."

"Sir!" said the Count Allamistakeo.

The Doctor repeated his remarks, but it was only after much additional explanation, that the foreigner could be made to comprehend them. The latter at length said, hesitatingly:

"The ideas you have suggested are to me, I confess, utterly novel. During my time I never knew any one to entertain so singular a fancy as that the universe (or this world if you will have it so) ever had a beginning at all. I remember once, and once only, hearing something remotely hinted, by a man of many speculations, concerning the origin *of the human race;* and by this individual, the very word *Adam* (or Red Earth), which you make use of, was employed. He employed it, however, in a generical sense, with reference to the spontaneous germination from rank soil (just as a thousand of the lower *genera* of creatures are germinated)— the spontaneous germination, I say, of five vast hordes of men, simultaneously upspringing in five distinct and nearly equal divisions of the globe."

Here, in general, the company shrugged their shoulders, and one or two of us touched our foreheads with a very significant air. Mr. Silk Buckingham, first glancing slightly at the occiput and then at the sinciput of Allamistakeo, spoke as follows:

"The long duration of human life in your time, together with the occasional practice of passing it, as you have explained, in instalments, must have had, indeed, a strong tendency to the general development and conglomeration of knowledge. I presume, therefore, that we are to attribute the marked inferiority of the old Egyptians in all particulars of science, when compared with the moderns, and more especially with the Yankees, altogether to the superior solidity of the Egyptian skull."

"I confess again," replied the Count, with much suavity, "that I am somewhat at a loss to comprehend you; pray, to what particulars of science do you allude?"

Here our whole party, joining voices, detailed, at great length, the assumptions of phrenology and the marvels of animal magnetism.

Having heard us to the end, the Count proceeded to relate a few anecdotes, which rendered it evident that prototypes of Gall and Spurzheim had flourished and faded in Egypt so long ago as to have been nearly forgotten, and that the manœuvres of Mesmer were really very contemptible tricks when put in collation with the positive miracles of the Theban *savants,* who created lice and a great many other similar things.

I here asked the Count if his people were able to calculate eclipses. He smiled rather contemptuously, and said they were.

This put me a little out, but I began to make other inquiries in regard to his astronomical knowledge, when a member of the company, who had never as yet opened his mouth, whispered in my ear, that for information on this head, I better consult Ptolemy, (whoever Ptolemy is) as well as one Plutarch *de facie lunæ.*

I then questioned the mummy about burning-glasses and lenses, and, in general, about the manufacture of glass; but I had not made an end of my queries before the silent member again touched me quietly on the elbow, and begged me for God's sake to take a peep at Diodorus Siculus. As for the Count, he merely asked me, in the way of reply, if we moderns possessed any such microscopes as would enable us to cut cameos in the style of the Egyptians. While I was thinking how I should answer this question, little Doctor Ponnonner committed himself in a very extraordinary way.

"Look at our architecture!" he exclaimed, greatly to the indignation of both the travelers, who pinched him black and blue to no purpose.

"Look," he cried with enthusiasm, "at the Bowling-Green Fountain in New York! or if this be too vast a contemplation, regard for a moment the Capitol at Washington, D.C.!"—and the good little medical man went on to detail very minutely the proportions of the fabric to which he referred. He explained that the portico alone was adorned with no less than four and twenty columns, five feet in diameter, and ten feet apart.

The Count said that he regretted not being able to remember, just at that moment, the precise dimensions of any one of the principal buildings of the city of Aznac, whose foundations were laid in the night of Time, but the ruins of which were still standing, at the epoch of his entombment, in a vast plain of sand to the westward of Thebes. He recollected, however, (talking of the porticoes) that one affixed to an inferior palace in a kind of suburb called Carnac, consisted of a hundred and forty-four columns, thirty-seven feet in circumference, and twenty-five feet apart. The approach to this portico, from the Nile, was through an avenue two miles long, composed of sphynxes, statues, and obelisks, twenty, sixty, and a hundred feet in height. The palace itself (as well as he could remember) was, in one direction, two miles long, and might have been altogether about seven in circuit. Its walls were richly painted all over, within and without, with hieroglyphics. He would not pretend to *assert* that even fifty or sixty of the Doctor's Capitols might have been built within these walls, but he was by no means sure that two or three hundred of them might not have been squeezed in with some trouble. That palace at Carnac was an insignificant little building after all. He, (the Count) however, could not conscientiously refuse to admit the ingenuity, magnificence, and superiority of the Fountain at the Bowling-Green, as described by the Doctor. Nothing like it, he was forced to allow, had ever been seen in Egypt or elsewhere.

I here asked the Count what he had to say to our rail-roads.

"Nothing," he replied, "in particular." They were rather slight, rather ill-conceived, and clumsily put together. They could not be compared, of course, with the vast, level, direct, iron-grooved causeways upon which the Egyptians conveyed entire temples and solid obelisks of a hundred and fifty feet in altitude.

I spoke of our gigantic mechanical forces.

He agreed that we knew something in that way, but inquired how I should have gone to work in getting up the imposts on the lintels of even the little palace at Carnac.

This question I concluded not to hear, and demanded if he had any idea of Artesian wells; but he simply raised his eye-brows; while Mr. Gliddon winked at me very hard and said, in a low tone, that one had been recently discovered by the engineers employed to bore for water in the Great Oasis.

I then mentioned our steel; but the foreigner elevated his nose, and asked me if our steel could have executed the sharp carved work seen on the obelisks, and which was wrought altogether by edge-tools of copper.

This disconcerted us so greatly that we thought it advisable to vary the attack to Metaphysics. We sent for a copy of a book called the "Dial," and read out of it a chapter or two about something which is not very clear, but which the Bostonians call the Great Movement of Progress.

The Count merely said that Great Movements were awfully common things in his day, and as for Progress, it was at one time quite a nuisance, but it never progressed.

We then spoke of the great beauty and importance of Democracy, and were at much trouble in impressing the Count with a due sense of the advantages we enjoyed in living where there was suffrage *ad libitum,* and no king.

He listened with marked interest, and in fact seemed not a little amused. When we had done, he said that, a great while ago, there had occurred something of a very similar sort. Thirteen Egyptian provinces determined all at once to be free, and to set a magnificent example to the rest of mankind. They assembled their wise men, and concocted the most ingenious constitution it is possible to conceive. For a while they managed remarkably well; only their habit of bragging was prodigious: The thing ended, however, in the consolidation of the thirteen states, with some fifteen or twenty others, in the most odious and insupportable despotism that was ever heard of upon the face of the Earth.

I asked what was the name of the usurping tyrant.

As well as the Count could recollect, it was *Mob.*

Not knowing what to say to this, I raised my voice, and deplored the Egyptian ignorance of steam.

The Count looked at me with much astonishment, but made no answer. The silent gentleman, however, gave me a violent nudge in the ribs with his elbows—told me I had sufficiently exposed myself for once—and demanded if I was really such a fool as not to know that the modern steam engine is derived from the invention of Hero, through Solomon de Caus.

We were now in imminent danger of being discomfited; but, as good luck would have it, Doctor Ponnonner, having rallied, returned to our rescue, and inquired if the people of Egypt would seriously pretend to rival the moderns in the all-important particular of dress.

The Count, at this, glanced downward to the straps of his pantaloons, and then taking hold of the end of one of his coat-tails, held it up close to his eyes for some minutes. Letting it fall, at last, his mouth extended itself very gradually from ear to ear; but I do not remember that he said anything in the way of reply.

Hereupon we recovered our spirits, and the Doctor, approaching the Mummy with great dignity, desired it to say candidly, upon its honor as a gentleman, if the Egyptians had comprehended, at *any* period, the manufacture of either Ponnonner's lozenges or Brandreth's pills.

We looked, with profound anxiety, for an answer;—but in vain. It was not forthcoming. The Egyptian blushed and hung down his head. Never was triumph more consummate; never was defeat borne with so ill a grace. Indeed, I could not endure the spectacle of the poor Mummy's mortification. I reached my hat, bowed to him stiffly, and took leave.

Upon getting home I found it past four o'clock, and went immediately to bed. It is now ten A.M. I have been up since seven, penning these memoranda for the benefit of my family and of mankind. The former I shall behold no more. My wife is a shrew. The truth is, I am heartily sick of this life and of the nineteenth century in general. I am convinced that every thing is going wrong. Besides, I am anxious to know who will be President in 2045. As soon, therefore, as I shave and swallow a cup of coffee, I shall just step over to Ponnonner's and get embalmed for a couple of hundred years.

DISCUSSION QUESTIONS

1. Who is the narrator, and what is his relationship to Doctor Ponnonner? What details from the story help characterize these men?

2. Can you imagine a contemporary analog for mummy-unwrapping parties? How, for example, does the anatomical study of cadavers compare to the type of study Doctor Pononner and associates attempt? In anatomy labs, what can be learned exclusively from the body that cannot be learned through other means? What cannot be learned from the body alone? What, for example, does the patient narrative add to the understanding of disease?

3. Poe uses scientific terminology and clinical imagery throughout the tale. What effect does the incorporation of this professional language have on the reader? Discuss instances in health care practice or biomedical research today in which the use of professional jargon might negatively affect communication and, ultimately, patient outcomes.

4. Why is the mummy disappointed in the two men who have travelled to Egypt? How does this tension introduce issues of cultural awareness and sensitivity? If the mummy had been a patient, how would his treatment have been complicated by the doctor's lack of this awareness and sensitivity? Why are these concepts considered important to improving patient outcomes today?

5. Unlike Poe's tales of horror, this is a humorous satire. What, then, is Poe satirizing? If Poe were living in our own century, could he poke fun at Americans for the same reasons? What would Poe think if he compared medico-scientific practices of the twenty-first century to those of the nineteenth century? Rewrite the tale, making Dr. Ponnonner and friends twenty-first century characters who awaken a nineteenth-century American man or woman.

6. The mummy's story indicates that the Egyptians had mastered many advanced technologies. Do our medico-scientific and technological advances in the twenty-first century ever threaten the delivery of quality health care even though they seem to promise more efficient methods of diagnosis and treatment? Reflect on a time when, as patient or provider, you felt technology interfered with your receipt or delivery of patient- or relationship-centered care. What steps could have been taken to ensure that technology supported instead of undermined care?

The System of Doctor Tarr and Professor Fether
November 1845, *Graham's Magazine*

How do kindness and compassion affect health outcomes? Our century has seen a resurgence of interest in this question as a growing body of evidence indicates that empathy and patient-centered care relate positively to improved patient outcomes. Yet, we are by far not the first to explore methods of improving the patient experience through humanistic means, by far not the first generation faced with the reality that the "art of medicine" often gets lost or obscured in a health care system in which labs, scans, medications, and advanced technologies are favored over the physical exam, the bedside manner, the patient narrative. In Poe's time, health care reformers sought to improve care of the mentally ill—and the disabled—who were increasingly outcast from communities, locked away in asylums. To this end, reformers promoted the "moral treatment" over the traditional method of confinement, endorsing humane treatment of mental health patients and basing care on the principles of "freedom and kindness" (Benton 1968, 126). By the early 1800s, American doctors had imported the method from Europe, and by the 1840s several asylums in Massachusetts, Connecticut, and New York had adopted the new system (Benton 1968, 126).

Years before Poe visited one of the asylums where the "moral treatment" or "soothing system" was in use, he had likely read about the system in Charles Dickens's 1843 American Notes *and Nathaniel Parker Willis's 1834 "The Madhouse of Palermo" (Levine and Levine 1990; Benton 1968). In fact, in this tale he may have been satirizing Dickens or Willis, because both men had offended him, and it was standard for Poe to lash out in print (Levine and Levine 1990). Although critics debate which man, if either, was the true target of his satire, Poe was clearly satirizing the method of reform by "problematizing the distinction between madness and sanity" upon which it rested (Kennedy 2001, 7).*

Written after Poe visited an asylum run by his acquaintance, Dr. Pliny Earle, resident physician and superintendent at asylums in Frankford, Pennsylvania, and Bloomingdale, New York (Poe 2008; Benton 1968), "The System of Doctor Tarr and Professor Fether" features characters living in a private mental health care institution in France, where patients are encouraged to roam freely and indulge every whim and fancy.[1] Poe invites readers to consider the results of this treatment if taken to the extremes. The French setting, however, does not detract from the commentary Poe makes on American reforms. While he seems to advance the idea that not all reform is good, he perhaps inadvertently causes readers to examine the process of change, to consider how systemic health care reform involves experimentation and adaptation that may lead to periods of uncertainty—even chaos—before order is restored.

•

DURING the autumn of 18—, while on a tour through the extreme Southern provinces of France, my route led me within a few miles of a certain *Maison de Santé* or private Mad-House, about which I had heard much, in Paris, from my medical friends. As I had never visited a place of the kind, I thought the opportunity too good to be lost; and so proposed to my traveling companion (a gentleman with whom I had made casual acquaintance, a few days before,) that we should turn aside, for an hour or so, and look through the establishment. To this he objected—pleading haste, in the first place, and, in the second, a very usual horror at the sight of a lunatic. He begged of me, however, not to let any mere courtesy toward himself interfere with the gratification of my curiosity, and said that he would ride on leisurely, so that I might overtake him during the day, or, at all events, during the next. As he bade me good-bye, I bethought me that there might be some difficulty in obtaining access to the premises, and mentioned my fears on this point. He replied that, in fact, unless I had personal knowledge of the superintendent, Monsieur Maillard, or some credential in the way of a letter, a difficulty might be found to exist, as the regulations of these private mad-houses were more rigid than the public hospital laws. For himself, he added, he had, some years since, made the acquaintance of Maillard, and would so far assist me as to ride up to the door and introduce me; although his feelings on the subject of lunacy would not permit of his entering the house.

I thanked him, and, turning from the main-road, we entered a grass-grown by-path, which, in half an hour, nearly lost itself in a dense forest, clothing the base of a mountain. Through this dank and gloomy wood we rode some two miles, when the *Maison de Santé* came in view. It was a fantastic *château*, much dilapidated, and indeed scarcely tenantable through age and neglect. Its aspect inspired me with absolute dread, and, checking my horse, I half resolved to turn back. I soon, however, grew ashamed of my weakness, and proceeded.

As we rode up to the gate-way, I perceived it slightly open, and the visage of a man peering through. In an instant afterward, this man came forth, accosted my companion by name, shook him cordially by the hand, and begged him to alight. It was Monsieur Maillard himself. He was a portly, fine-looking gentleman of the old school, with a polished manner, and a certain air of gravity, dignity, and authority which was very impressive.

My friend, having presented me, mentioned my desire to inspect the establishment, and received Monsieur Maillard's assurance that he would show me all attention, now took leave, and I saw him no more.

When he had gone, the superintendent ushered me into a small and exceedingly neat parlor, containing, among other indications of refined taste, many books, drawings, pots of flowers, and musical instruments. A cheerful fire blazed upon the hearth. At a piano, singing an aria from Bellini, sat a young and very beautiful woman, who, at my entrance, paused in her song, and received me with graceful courtesy. Her voice was low, and her whole manner subdued. I thought, too, that I perceived the traces of sorrow in her countenance, which was excessively, although, to my taste, not unpleasingly pale. She was attired in deep mourning, and excited in my bosom a feeling of mingled respect, interest, and admiration.

I had heard, at Paris, that the institution of Monsieur Maillard was managed upon what is vulgarly termed the "system of soothing"—that all punishments were avoided—that even confinement was seldom resorted to—that the patients, while secretly watched, were left much apparent liberty, and that most of them were permitted to roam about the house and grounds in the ordinary apparel of persons in right mind.

Keeping these impressions in view, I was cautious in what I said before the young lady; for I could not be sure that she was sane; and, in fact, there was a certain restless brilliancy about her eyes which half led me to imagine she was not. I confined my remarks, therefore, to general topics, and to such as I thought would not be displeasing or exciting even to a lunatic. She replied in a perfectly rational manner to all that I said; and even her original observations were marked with the soundest good sense; but a long acquaintance with the metaphysics of *mania,* had taught me to put no faith in such evidence of sanity, and I continued to practice, throughout the interview, the caution with which I commenced it.

Presently a smart footman in livery brought in a tray with fruit, wine, and other refreshments, of which I partook, the lady soon afterward leaving the room. As she departed I turned my eyes in an inquiring manner toward my host.

"No," he said, "oh, no—a member of my family—my niece, and a most accomplished woman."

"I beg a thousand pardons for the suspicion," I replied, "but of course you will know how to excuse me. The excellent administration of your affairs here is well understood in Paris, and I thought it just possible, you know—"

"Yes, yes—say no more—or rather it is myself who should thank you for the commendable prudence you have displayed. We seldom find so much of forethought in young men; and, more than once, some unhappy *contre-temps* has occurred in consequence of thoughtlessness on the part of our visiters. While my former system was in operation, and my patients were permitted the privilege of roaming to and fro at will, they were often aroused to a dangerous frenzy by injudicious persons who called to inspect the house. Hence I was obliged to enforce a rigid system of exclusion; and none obtained access to the premises upon whose discretion I could not rely."

"While your *former* system was in operation!" I said, repeating his words— "do I understand you, then, to say that the 'soothing system' of which I have heard so much is no longer in force?"

"It is now," he replied, "several weeks since we have concluded to renounce it forever."

"Indeed! you astonish me!"

"We found it, sir," he said, with a sigh, "absolutely necessary to return to the old usages. The *danger* of the soothing system was, at all times, appalling; and its advantages have been much overrated. I believe, sir, that in this house it has been given a fair trial, if ever in any. We did every thing that rational humanity could suggest. I am sorry that you could not have paid us a visit at an earlier period, that you might have judged for yourself. But I presume you are conversant with the soothing practice—with its details."

"Not altogether. What I have heard has been at third or fourth hand."

"I may state the system then, in general terms, as one in which the patients were *menagés*—humored. We contradicted *no* fancies which entered the brains of the mad. On the contrary, we not only indulged but encouraged them; and many of our most permanent cures have been thus effected. There is no argument which so touches the feeble reason of the madman as the *argumentum ad absurdum*. We have had men, for example, who fancied themselves chickens. The cure was, to insist upon the thing as a fact—to accuse the patient of stupidity in not sufficiently perceiving it to be a fact—and thus to refuse him any other diet for a week than that which properly appertains to a chicken. In this manner a little corn and gravel were made to perform wonders."

"But was this species of acquiescence all?"

"By no means. We put much faith in amusements of a simple kind, such as music, dancing, gymnastic exercises generally, cards, certain classes of books, and so forth. We affected to treat each individual as if for some ordinary physical disorder; and the word 'lunacy' was never employed. A great point was to set each lunatic to guard the actions of all the others. To repose confidence in the understanding or discretion of a madman, is to gain him body and soul. In this way we were enabled to dispense with an expensive body of keepers."

"And you had no punishments of any kind?"

"None."

"And you never confined your patients?"

"Very rarely. Now and then, the malady of some individual growing to a crisis, or taking a sudden turn of fury, we conveyed him to a secret cell, lest his disorder should infect the rest, and there kept him until we could dismiss him to his friends—for with the raging maniac we have nothing to do. He is usually removed to the public hospitals."

"And you have now changed all this—and you think for the better?"

"Decidedly. The system had its disadvantages, and even its dangers. It is now, happily, exploded throughout all the *Maisons de Santé* of France."

"I am very much surprised," I said, "at what you tell me; for I made sure that, at this moment, no other method of treatment for mania existed in any portion of the country."

"You are young yet, my friend," replied my host, "but the time will arrive when you will learn to judge for yourself of what is going on in the world, without trusting to the gossip of others. Believe nothing you hear, and only one half that you see. Now about our *Maisons de Santé*, it is clear that some ignoramus has misled you. After dinner, however, when you have sufficiently recovered from the fatigue of your ride, I will be happy to take you over the house, and introduce to you a system which, in my opinion, and in that of every one who has witnessed its operation, is incomparably the most effectual as yet devised."

"Your own?" I inquired—"one of your own invention?"

"I am proud," he replied, "to acknowledge that it is—at least in some measure."

In this manner I conversed with Monsieur Maillard for an hour or two, during which he showed me the gardens and conservatories of the place.

"I cannot let you see my patients," he said, "just at present. To a sensitive mind there is always more or less of the shocking in such exhibitions; and I do not wish to spoil your appetite for dinner. We will dine. I can give you some veal *á la St. Menehoult,* with cauliflowers in *velouté* sauce—after that a glass of *Clos de Vougeot*—then your nerves will be sufficiently steadied."

At six, dinner was announced; and my host conducted me into a large *salle à manger,* where a very numerous company were assembled—twenty-five or thirty in all. They were, apparently, people of rank—certainly of high breed-ing—although their habiliments, I thought, were extravagantly rich, partaking somewhat too much of the ostentatious finery of the *vieille cour.* I noticed that at least two-thirds of these guests were ladies; and some of the latter were by no means accoutred in what a Parisian would consider good taste at the present day. Many females, for example, whose age could not have been less than seventy, were bedecked with a profusion of jewelry, such as rings, bracelets, and ear-rings, and wore their bosoms and arms shamefully bare. I observed, too, that very few of the dresses were well made—or, at least, that very few of them fitted the wearers. In looking about, I discovered the interesting girl to whom Monsieur Maillard had presented me in the little parlor; but my surprise was great to see her wearing a hoop and farthingale, with high-heeled shoes, and a dirty cap of Brussels lace, so much too large for her that it gave her face a ridiculously diminutive expression. When I had first seen her, she was attired, most becomingly, in deep mourning. There was an air of oddity, in short, about the dress of the whole party, which, at first, caused me to recur to my original idea of the "soothing system," and to fancy that Monsieur Maillard had been willing to deceive me until after dinner, that I might experience no uncomfortable feelings during the repast, at finding myself dining with lunatics; but I remembered having been informed, in Paris, that the southern provincialists were a peculiarly eccentric people, with a vast number of antiquated notions; and then, too, upon conversing with several members of the company, my apprehensions were immediately and fully dispelled.

The dining-room itself, although perhaps sufficiently comfortable and of good dimensions, had nothing too much of elegance about it. For example, the floor was uncarpeted; in France, however, a carpet is frequently dispensed with. The windows, too, were without curtains; the shutters, being shut, were securely fastened with iron bars, applied diagonally, after the fashion of our ordinary shop-shutters. The apartment, I observed, formed, in itself, a wing of the *château,* and thus the windows were on three sides of the parallelogram, the door being at the other. There were no less than ten windows in all.

The table was superbly set out. It was loaded with plate, and more than loaded with delicacies. The profusion was absolutely barbaric. There were meats enough to have feasted the Anakim. Never, in all my life, had I witnessed so lavish, so wasteful an expenditure of the good things of life. There seemed very little taste, however, in the arrangements; and my eyes, accustomed to quiet lights, were sadly offended by the prodigious glare of a multitude of wax candles, which, in silver *candelabra,* were deposited upon the table, and all about the room, wherever it was possible to find a place. There were several active servants in attendance; and, upon a large table, at the farther end of the apartment, were seated seven or eight people with fiddles, fifes, trombones, and a drum. These fellows annoyed me very much, at intervals, during the repast, by an infinite variety of noises, which were intended for music, and which appeared to afford much entertainment to all present, with the exception of myself.

Upon the whole, I could not help thinking that there was much of the *bizarre* about every thing I saw—but then the world is made up of all kinds of persons, with all modes of thought, and all sorts of conventional customs. I had traveled, too, so much, as to be quite an adept at the *nil admirari,* so I took my seat very coolly at the right hand of my host, and, having an excellent appetite, did justice to the good cheer set before me.

The conversation, in the mean time, was spirited and general. The ladies, as usual, talked a great deal. I soon found that nearly all the company were well educated; and my host was a world of good-humored anecdote in himself. He seemed quite willing to speak of his position as superintendent of a *Maison de Santé;* and, indeed, the topic of lunacy was, much to my surprise, a favorite one with all present. A great many amusing stories were told, having reference to the *whims* of the patients.

"We had a fellow here once," said a fat little gentleman, who sat at my right,—"a fellow that fancied himself a tea-pot; and, by the way, is it not especially singular how often this particular crotchet has entered the brain of the lunatic? There is scarcely an insane asylum in France which cannot supply a human tea-pot. *Our* gentleman was a Britannia-ware tea-pot, and was careful to polish himself every morning with buckskin and whiting."

"And then," said a tall man just opposite, "we had here, not long ago, a person who had taken it into his head that he was a donkey—which, allegorically speaking, you will say, was quite true. He was a troublesome patient; and we had much ado to keep him within bounds. For a long time he would eat nothing but thistles; but of this idea we soon cured him by insisting upon his eating nothing else. Then he was perpetually kicking out his heels—so—so—"

"Mr. De Kock! I will thank you to behave yourself!" here interrupted an old lady, who sat next to the speaker. "Please keep your feet to yourself! You have spoiled my brocade! Is it necessary, pray, to illustrate a remark in so practical a style? Our friend here can surely comprehend you without all this. Upon my word, you are nearly as great a donkey as the poor unfortunate imagined himself. Your acting is very natural, as I live!"

"Mille pardons! mam'selle!" replied Monsieur De Kock, thus addressed—"a thousand pardons! I had no intention of offending. Mam'selle Laplace—Monsieur De Kock will do himself the honor of taking wine with you."

Here Monsieur De Kock bowed low, kissed his hand with much ceremony, and took wine with Mam'selle Laplace.

"Allow me, *mon ami*," now said Monsieur Maillard, addressing myself, "allow me to send you a morsel of this veal *à la St. Menehoult*—you will find it particularly fine."

At this instant three sturdy waiters had just succeeded in depositing safely upon the table an enormous dish, or trencher, containing what I supposed to be the "*monstrum, horrendum, injorme, ingens, cui lumen ademptum.*" A closer scrutiny assured me, however, that it was only a small calf roasted whole, and set upon its knees, with an apple in its mouth, as is the English fashion of dressing a hare.

"Thank you, no," I replied; "to say the truth, I am not particularly partial to veal *à la St.*—what is it?—for I do not find that it altogether agrees with me. I will change my plate, however, and try some of the rabbit."

There were several side-dishes on the table, containing what appeared to be the ordinary French rabbit—a very delicious *morceau,* which I can recommend.

"Pierre," cried the host, "change this gentleman's plate, and give him a side-piece of this rabbit *au-chat.*"

"This what?" said I.

"This rabbit *au-chat.*"

"Why, thank you—upon second thoughts, no. I will just help myself to some of the ham."

There is no knowing what one eats, thought I to myself, at the tables of these people of the province. I will have none of their rabbit *au-chat*—and, for the matter of that, none of their *cat-au-rabbit* either.

"And then," said a cadaverous looking personage, near the foot of the table, taking up the thread of the conversation where it had been broken off—"and then, among other oddities, we had a patient, once upon a time, who very pertinaciously maintained himself to be a Cordova cheese, and went about, with a knife in his hand, soliciting his friends to try a small slice from the middle of his leg."

"He was a great fool, beyond doubt," interposed some one, "but not to be compared with a certain individual whom we all know, with the exception of this strange gentleman. I mean the man who took himself for a bottle of champagne, and always went off with a pop and a fizz, in this fashion."

Here the speaker, very rudely, as I thought, put his right thumb in his left cheek, withdrew it with a sound resembling the popping of a cork, and then, by a dexterous movement of the tongue upon the teeth, created a sharp hissing and fizzing, which lasted for several minutes, in imitation of the frothing of champagne. This behavior, I saw plainly, was not very pleasing to Monsieur

Maillard; but that gentleman said nothing, and the conversation was resumed by a very lean little man in a big wig.

"And then there was an ignoramus," said he, "who mistook himself for a frog; which, by the way, he resembled in no little degree. I wish you could have seen him, sir"—here the speaker addressed myself—"it would have done your heart good to see the natural airs that he put on. Sir, if that man was *not* a frog, I can only observe that it is a pity he was not. His croak thus—o-o-o-o-gh—o-o-o-o-gh! was the finest note in the world—B flat; and when he put his elbows upon the table thus—after taking a glass or two of wine—and distended his mouth, thus, and rolled up his eyes, thus, and winked them with excessive rapidity, thus, why then, sir, I take it upon myself to say, positively, that you would have been lost in admiration of the genius of the man."

"I have no doubt of it," I said.

"And then," said somebody else, "then there was Petit Gaillard, who thought himself a pinch of snuff, and was truly distressed because he could not take himself between his own finger and thumb."

"And then there was Jules Desoulières, who was a very singular genius, indeed, and went mad with the idea that he was a pumpkin. He persecuted the cook to make him up into pies—a thing which the cook indignantly refused to do. For my part, I am by no means sure that a pumpkin pie *à la Desoulières,* would not have been very capital eating indeed!"

"You astonish me!" said I; and I looked inquisitively at Monsieur Maillard.

"Ha! ha! ha!" said that gentleman—"he! he! he!—hi! hi! hi!—ho! ho! ho!—hu! hu! hu!—very good indeed! You must not be astonished, *mon ami;* our friend here is a wit—a *drôle*—you must not understand him to the letter."

"And then," said some other one of the party, "then there was Bouffon Le Grand—another extraordinary personage in his way. He grew deranged through love, and fancied himself possessed of two heads. One of these he maintained to be the head of Cicero; the other he imagined a composite one, being Demosthenes' from the top of the forehead to the mouth, and Lord Brougham from the mouth to the chin. It is not impossible that he was wrong; but he would have convinced you of his being in the right; for he was a man of great eloquence. He had an absolute passion for oratory, and could not refrain from display. For example, he used to leap upon the dinner-table thus, and—and—"

Here a friend, at the side of the speaker, put a hand upon his shoulder and whispered a few words in his ear; upon which he ceased talking with great suddenness, and sank back within his chair.

"And then," said the friend who had whispered, "there was Boullard, the tee-totum. I call him the tee-totum, because, in fact, he was seized with the droll, but not altogether irrational crotchet, that he had been converted into a tee-totum. You would have roared with laughter to see him spin. He would turn round upon one heel by the hour, in this manner—so—"

Here the friend whom he had just interrupted by a whisper, performed an exactly similar office for himself.

"But then," cried the old lady, at the top of her voice, "your Monsieur Boullard was a madman, and a very silly madman at best; for who, allow me to ask you, ever heard of a human tee-totum? The thing is absurd. Madame Joyeuse was a more sensible person, as you know. She had a crotchet, but it was instinct with common sense, and gave pleasure to all who had the honor of her acquaintance. She found, upon mature deliberation, that, by some accident, she had been turned into a chicken-cock; but, as such, she behaved with propriety. She flapped her wings with prodigious effect—so—so—so—and, as for her crow, it was delicious! Cock-a-doodle-doo!—cock-a-doodle-doo!—cock-a-doodle-de-doo-doo-dooo-do-o-o-o-o-o-o-!"

"Madame Joyeuse, I will thank you to behave yourself!" here interrupted our host, very angrily. "You can either conduct yourself as a lady should do, or you can quit the table forthwith—take your choice."

The lady (whom I was much astonished to hear addressed as Madame Joyeuse, after the description of Madame Joyeuse she had just given,) blushed up to the eye-brows, and seemed exceedingly abashed at the reproof. She hung down her head, and said not a syllable in reply. But another and younger lady resumed the theme. It was my beautiful girl of the little parlor!

"Oh, Madame Joyeuse *was* a fool!" she exclaimed; "but there was really much sound sense, after all, in the opinion of Eugénie Salsafette. She was a very beautiful and painfully modest young lady, who thought the ordinary mode of habiliment indecent, and wished to dress herself, always, by getting outside instead of inside of her clothes. It is a thing very easily done, after all. You have only to do so—and then so—so—so—and then so—so—so—and then—"

"Mon dieu! Mam'selle Salsafette!" here cried a dozen voices at once. "What *are* you about?—forbear!—that is sufficient!—we see, very plainly, how it is done!—hold! hold!" and several persons were already leaping from their seats to withhold Mam'selle Salsafette from putting herself upon a par with the Medicean Venus, when the point was very effectually and suddenly accomplished by a series of loud screams, or yells, from some portion of the main body of the château.

My nerves were very much affected, indeed, by these yells; but the rest of the company I really pitied. I never saw any set of reasonable people so thoroughly frightened in my life. They all grew as pale as so many corpses, and, shrinking within their seats, sat quivering and gibbering with terror, and listening for a repetition of the sound. It came again—louder and seemingly nearer—and then a third time *very* loud, and then a fourth time with a vigor evidently diminished. At this apparent dying away of the noise, the spirits of the company were immediately regained, and all was life and anecdote as before. I now ventured to inquire the cause of the disturbance.

"A mere *bagatelle*," said Monsieur Maillard. "We are used to these things, and care really very little about them. The lunatics, every now and then, get up a howl in concert; one starting another, as is sometimes the case with a bevy of dogs at night. It occasionally happens, however, that the *concerto* yells are succeeded by a simultaneous effort at breaking loose; when, of course, some little danger is to be apprehended."

"And how many have you in charge?"

"At present we have not more than ten, altogether."

"Principally females, I presume?"

"Oh, no—every one of them men, and stout fellows, too, I can tell you."

"Indeed! I have always understood that the majority of lunatics were of the gentler sex."

"It is generally so, but not always. Some time ago, there were about twenty-seven patients here; and, of that number, no less than eighteen were women; but, lately, matters have changed very much, as you see."

"Yes—have changed very much, as you see," here interrupted the gentleman who had broken the shins of Mam'selle Laplace.

"Yes—have changed very much, as you see!" chimed in the whole company at once.

"Hold your tongues, every one of you!" said my host, in a great rage. Whereupon the whole company maintained a dead silence for nearly a minute. As for one lady, she obeyed Monsieur Maillard to the letter, and thrusting out her tongue, which was an excessively long one, held it very resignedly, with both hands, until the end of the entertainment.

"And this gentlewoman," said I, to Monsieur Maillard, bending over and addressing him in a whisper—"this good lady who has just spoken, and who gives us the cock-a-doodle-de-doo—she, I presume, is harmless—quite harmless, eh?"

"Harmless!" ejaculated he, in unfeigned surprise, "why—why, what *can* you mean?"

"Only slightly touched?" said I, touching my head. "I take it for granted that she is not particularly—not dangerously affected, eh?"

"*Mon Dieu!* what *is* it you imagine? This lady, my particular old friend, Madame Joyeuse, is as absolutely sane as myself. She has her little eccentricities, to be sure—but then, you know, all old women—all *very* old women—are more or less eccentric!"

"To be sure," said I,—"to be sure—and then the rest of these ladies and gentlemen—"

"Are my friends and keepers," interrupted Monsieur Maillard, drawing himself up with *hauteur*,—"my very good friends and assistants."

"What! all of them?" I asked—"the women and all?"

"Assuredly," he said,—"we could not do at all without the women; they are the best lunatic-nurses in the world; they have a way of their own, you know; their bright eyes have a marvellous effect;—something like the fascination of the snake, you know."

"To be sure," said I,—"to be sure! They behave a little odd, eh?—they are a little *queer*, eh?—don't you think so?"

"Odd!—queer!—why, do you *really* think so? We are not very prudish, to be sure, here in the South—do pretty much as we please—enjoy life, and all that sort of thing, you know—"

"To be sure," said I,—"to be sure."

"And then, perhaps, this *Clos de Vougeot* is a little heady, you know—a little *strong*—you understand, eh?"

"To be sure," said I,—"to be sure. By-the-bye, monsieur, did I understand you to say that the system you have adopted, in place of the celebrated soothing system, was one of very rigorous severity?"

"By no means. Our confinement is necessarily close; but the treatment—the medical treatment, I mean—is rather agreeable to the patients than otherwise."

"And the new system is one of your own invention?"

"Not altogether. Some portions of it are referable to Professor Tarr, of whom you have, necessarily, heard; and, again, there are modifications in my plan which I am happy to acknowledge as belonging of right to the celebrated Fether, with whom, if I mistake not, you have the honor of an intimate acquaintance."

"I am quite ashamed to confess," I replied, "that I have never even heard the names of either gentleman before."

"Good heavens!" ejaculated my host, drawing back his chair abruptly, and uplifting his hands. "I surely do not hear you aright! You did not intend to say, eh? that you had never *heard* either of the learned Doctor Tarr, or of the celebrated Professor Fether?"

"I am forced to acknowledge my ignorance," I replied; "but the truth should be held inviolate above all things. Nevertheless, I feel humbled to the dust, not to be acquainted with the works of these no doubt extraordinary men. I will seek out their writings forthwith, and peruse them with deliberate care.

Monsieur Maillard, you have really—I must confess it—you have *really*—made me ashamed of myself!"

And this was the fact.

"Say no more, my good young friend," he said kindly, pressing my hand,—"join me now in a glass of Sauterne."

We drank. The company followed our example without stint. They chatted—they jested—they laughed—they perpetrated a thousand absurdities—the fiddles shrieked—the drum row-de-dowed—the trombones bellowed like so many brazen bulls of Phalaris—and the whole scene, growing gradually worse and worse, as the wines gained the ascendancy, became at length a sort of pandemonium *in petto.* In the meantime, Monsieur Maillard and myself, with some bottles of Sauterne and Vougeot between us, continued our conversation at the top of the voice. A word spoken in an ordinary key stood no more chance of being heard than the voice of a fish from the bottom of Niagara Falls.

"And, sir," said I, screaming in his ear, "you mentioned something before dinner, about the danger incurred in the old system of soothing. How is that?"

"Yes," he replied, "there was, occasionally, very great danger, indeed. There is no accounting for the caprices of madmen; and, in my opinion, as well as in that of Doctor Tarr and Professor Fether, it is *never* safe to permit them to run at large unattended. A lunatic may be 'soothed,' as it is called, for a time, but, in the end, he is very apt to become obstreperous. His cunning, too, is proverbial, and great. If he has a project in view, he conceals his design with a marvellous wisdom; and the dexterity with which he counterfeits sanity, presents, to the metaphysician, one of the most singular problems in the study of mind. When a madman appears *thoroughly* sane, indeed, it is high time to put him in a strait-jacket."

"But the *danger,* my dear sir, of which you were speaking—in your own experience—during your control of this house—have you had practical reason to think liberty hazardous, in the case of a lunatic?"

"Here?—in my own experience?—why, I may say, yes. For example:—no *very* long while ago, a singular circumstance occurred in this very house. The 'soothing system,' you know, was then in operation, and the patients were at large. They behaved remarkably well—especially so—any one of sense might have known that some devilish scheme was brewing from that particular fact, that the fellows behaved so *remarkably* well. And, sure enough, one fine morning the keepers found themselves pinioned hand and foot, and thrown into the cells, where they were attended, as if *they* were the lunatics, by the lunatics themselves, who had usurped the offices of the keepers."

"You don't tell me so! I never heard of anything so absurd in my life!"

"Fact—it all came to pass by means of a stupid fellow—a lunatic—who, by some means, had taken it into his head that he had invented a better system of government than any ever heard of before—of lunatic government, I mean. He wished to give his invention a trial, I suppose—and so he persuaded the rest of the patients to join him in a conspiracy for the overthrow of the reigning powers."

"And he really succeeded?"

"No doubt of it. The keepers and kept were soon made to exchange places. Not that exactly either—for the madmen had been free, but the keepers were shut up in cells forthwith, and treated, I am sorry to say, in a very cavalier manner."

"But I presume a counter revolution was soon effected. This condition of things could not have long existed. The country people in the neighborhood—visiters coming to see the establishment—would have given the alarm."

"There you are out. The head rebel was too cunning for that. He admitted no visiters at all—with the exception, one day, of a very stupid-looking young gentleman of whom he had no reason to be afraid. He let him in to see the place—just by way of variety—to have a little fun with him. As soon as he had gammoned him sufficiently, he let him out, and sent him about his business."

"And *how* long, then, did the madmen reign?"

"Oh, a very long time, indeed—a month certainly—how much longer I can't precisely say. In the mean time, the lunatics had a jolly season of it—that you may swear. They doffed their own shabby clothes, and made free with the family wardrobe and jewels. The cellars of the *château* were well stocked with wine; and these madmen are just the devils that know how to drink it. They lived well, I can tell you."

"And the treatment—what was the particular species of treatment which the leader of the rebels put into operation?"

"Why, as for that, a madman is not necessarily a fool, as I have already observed; and it is my honest opinion that his treatment was a much better treatment than that which it superseded. It was a very capital system indeed—simple—neat—no trouble at all—in fact it was delicious—it was—"

Here my host's observations were cut short by another series of yells, of the same character as those which had previously disconcerted us. This time, however, they seemed to proceed from persons rapidly approaching.

"Gracious Heavens!" I ejaculated—"the lunatics have most undoubtedly broken loose."

"I very much fear it is so," replied Monsieur Maillard, now becoming excessively pale. He had scarcely finished the sentence, before loud shouts and imprecations were heard beneath the windows; and, immediately afterward, it became evident that some persons outside were endeavoring to gain entrance into the room. The door was beaten with what appeared to be a sledge-hammer, and the shutters were wrenched and shaken with prodigious violence.

A scene of the most terrible confusion ensued. Monsieur Maillard, to my excessive astonishment, threw himself under the side-board. I had expected more resolution at his hands. The members of the orchestra, who, for the last fifteen minutes, had been seemingly too much intoxicated to do duty, now sprang all at once to their feet and to their instruments, and, scrambling upon their table, broke out, with one accord, into "Yankee Doodle," which they performed, if not exactly in tune, at least with an energy superhuman, during the whole of the uproar.

Meantime, upon the main dining-table, among the bottles and glasses, leaped the gentleman who, with such difficulty, had been restrained from leaping there before. As soon as he fairly settled himself, he commenced an oration, which, no doubt, was a very capital one, if it could only have been heard. At the same moment, the man with the tee-totum predilection, set himself to spinning around the apartment, with immense energy, and with arms outstretched at right angles with his body; so that he had all the air of a tee-totum in fact, and knocked every body down that happened to get in his way. And now, too, hearing an incredible popping and fizzing of champagne, I discovered, at length, that it proceeded from the person who performed the bottle of that delicate drink during dinner. And then, again, the frog-man croaked away as if the salvation of his soul depended upon every note that he uttered. And, in the midst of all this, the continuous braying of a donkey arose over all. As for my old friend, Madame Joyeuse, I really could have wept for the poor lady, she appeared so terribly perplexed. All she did, however, was to stand up in a corner, by the fire-place, and sing out incessantly at the top of her voice, "Cock-a-doodle-de-doooooooh!"

And now came the climax—the catastrophe of the drama. As no resistance, beyond whooping and yelling and cock-a-doodleing, was offered to the en-croachments of the party without, the ten windows were very speedily, and almost simultaneously, broken in. But I shall never forget the emotions of wonder and horror with which I gazed, when, leaping through these windows, and down among us *pêle-mêle,* fighting, stamping, scratching, and howling, there rushed a perfect army of what I took to be chimpanzees, Ourang-Outangs, or big black baboons of the Cape of Good Hope.

I received a terrible beating—after which I rolled under a sofa, and lay still. After lying there some fifteen minutes, however, during which time I listened with all my ears to what was going on in the room, I came to some satisfactory *dénouement* of this tragedy. Monsieur Maillard, it appeared, in giving me the account of the lunatic who had excited his fellows to rebellion, had been merely relating his own exploits. This gentleman had, indeed, some two or three years before, been the superintendent of the establishment; but grew crazy himself, and so became a patient. This fact was unknown to the travelling companion who introduced me. The keepers, ten in number, having been suddenly over-powered, were first well tarred, then carefully feathered, and then shut up in underground cells. They had been so imprisoned for more than a month, during which period Monsieur Maillard had generously allowed them not only the tar and feathers (which constituted his "system") but some bread, and abundance of water. The latter was pumped on them daily. At length, one escaping through a sewer, gave freedom to all the rest.

The "soothing system," with important modifications, has been resumed at the *château;* yet I cannot help agreeing with Monsieur Maillard, that his own "treatment" was a very capital one of its kind. As he justly observed, it was "simple—neat—and gave no trouble at all—not the least."

I have only to add that, although I have searched every library in Europe for the works of Doctor *Tarr* and Professor *Fether,* I have, up to the present day, utterly failed in my endeavors at procuring an edition.

DISCUSSION QUESTIONS

1. Describe the mental health care facility that the narrator visits. What criticisms of the "soothing system" are implicit in this tale? Also, what does Poe seem to imply about the difference between the sane and insane in this context?

2. In what ways does the tale handle stereotypes about gender, class, and mental illness? What assumptions, for example, has the narrator made about the type of person who typically suffers from mental illness? How do socio-cultural norms shape the definitions of sanity and insanity, and how might these definitions disproportionately affect minority, marginalized, or otherwise disenfranchised groups?

3. This tale introduces issues related to patient authority and autonomy. Does the concept of patient autonomy function differently in the mental health care system than in the health care system at large? Provide examples to support your argument.

4. Watch the film *One Flew Over the Cuckoo's Nest* (1975) or *Shutter Island* (2010) and compare its plot and themes to those in "The System of Doctor Tarr and Professor Fether." Describe the ways each work reflects timeless issues related to mental health and mental health care treatments as well as issues specific to the eras in which they were created.

5. Today, patients seek out mental health care through the Internet, which introduces a number of issues about access to and quality of care. What other mental health care reforms and changes are occurring today thanks to new technologies? Speculate on what mental health care might look like in the future. Draw from Poe's story and at least one outside source that examines current practices.

6. Some scholars have seen in Poe's tale analogies to the system of slavery in antebellum America. Research the types of health care available to slaves on southern plantations, from African folk-healers to allopathic physicians. In what ways did the economic industry of slavery affect the way doctors viewed, treated, and experimented upon enslaved men and women, their health, and their bodies?

The Facts in the Case of M. Valdemar

December 1845, *American Whig Review*

and *Broadway Journal*

Poe biographer Jeffrey Meyers (1992) has called "The Facts in the Case of M. Valde-mar," the third and final story in Poe's mesmerism series, the "most extreme—and most effective—example of Poe's use of pathological details to convey the power of disease"' (178–79). Poe's fascination with the disease process is evident in his review of a book on abdominal disease, in which he wrote, "the pathology of fever in general has been at all times a fruitful subject of discussion" (178–79). Long (1989) notes the rise of the "anatomical-clinical" (22) practice of medicine in the nineteenth century gave Poe a new language to use in his writing, arguing that Poe adopts the genre of the "autopsy report" to describe the patient's condition in this story about "deranged scientific power brought to bear on death and disease" (33).

Poe's adoption of medico-scientific language and his expert hoaxing caused many readers to take the story as fact, not fiction (Thomas and Jackson 1987). This confusion was caused in part because the story was reprinted under titles like "Mesmerism in America. Death of M. Valdemar of New York" and "Mesmerism in America. Astounding and Horrifying Narrative" (615). In an era when tales and nonfiction pieces were often published side by side with no clear delineation between the two, these titles sounded to many readers like those of medical case studies. Poe received letters from the United States and beyond—from average gentleman to physicians and druggists—inquiring about the authenticity of the case (Thomas and Jackson 1987). Some were skeptical, like the editors of the Morning Post, *who noted that there were "several statements made, more especially with regard to the disease of which the patient died, which at once prove the case to be either a fabrication, or the work of one little acquainted with consumption. The story, however, is wonderful" (615). The* Popular Record of Modern Science *reprinted the article from the* Morning Post *with the same disclaimer about its potential fabrication. However, the editors of the* Record *believed the story a credible one, and they were dedicated to finding more "evidence upon the subject," including the identities of the doctors and medical students featured in the story. One reader of the* Broadway Journal *wrote that he admired the story either way: "The article suits me very well whether it is fact or fiction" (615).*

In this tale, the mesmerist extends the life of a dying patient unnaturally. The patient provides what Altschuler (2003) claims is one of the earliest examples of informed consent in literature.[1] Witnessing the event are two doctors, two nurses, and a medical student, who willingly allow the mesmerist to place himself at the top of their interprofessional hierarchy. The balance (or imbalance) of power among these providers—and the once-empowered patient—offers an interesting study in how professional roles are understood today compared to in the nineteenth century. More generally, the story prompts conversations about end-of-life issues, including pain management, palliative care, and patient autonomy, and inspires questions about the relationship between bench and bedside and the moral intricacies of using human subjects in medical research.

•

OF course I shall not pretend to consider it any matter for wonder, that the extraordinary case of M. Valdemar has excited discussion. It would have been a miracle had it not—especially under the circumstances. Through the desire of all parties concerned, to keep the affair from the public, at least for the present, or until we had further opportunities for investigation—through our endeavors to effect this—a garbled or exaggerated account made its way into society, and became the source of many unpleasant misrepresentations, and, very naturally, of a great deal of disbelief.

It is now rendered necessary that I give the *facts*—as far as I comprehend them myself. They are, succinctly, these:

My attention, for the last three years, had been repeatedly drawn to the subject of Mesmerism; and, about nine months ago, it occurred to me, quite suddenly, that in the series of experiments made hitherto, there had been a very remarkable and most unaccountable omission:—no person had as yet been mesmerized in *articulo mortis*. It remained to be seen, first, whether, in such condition, there existed in the patient any susceptibility to the magnetic influence; secondly, whether, if any existed, it was impaired or increased by the condition; thirdly, to what extent, or for how long a period, the encroachments of Death might be arrested by the process. There were other points to be ascertained, but these most excited my curiosity—the last in especial, from the immensely important character of its consequences.

In looking around me for some subject by whose means I might test these particulars, I was brought to think of my friend, M. Ernest Valdemar, the well-known compiler of the "Bibliotheca Forensica," and author (under the *nom de plume* of Issachar Marx) of the Polish versions of "Wallenstein" and "Gargantua." M. Valdemar, who has resided principally at Harlaem, N.Y., since the year of 1839, is (or was) particularly noticeable for the extreme spareness of his person—his lower limbs much resembling those of John Randolph; and, also, for the whiteness of his whiskers, in violent contrast to the blackness of his hair—the latter, in consequence, being very generally mistaken for a wig. His temperament was markedly nervous, and rendered him a good subject for mesmeric experiment. On two or three occasions I had put him to sleep with little difficulty, but was disappointed in other results which his peculiar constitution had naturally led me to anticipate. His will was at no period positively, or thoroughly, under my control, and in regard to *clairvoyance,* I could accomplish with him nothing to be relied upon. I always attributed my failure at these points to the disordered state of his health. For some months previous to my becoming acquainted with him, his physicians had declared him in a confirmed phthisis. It was his custom, indeed, to speak calmly of his approaching dissolution, as of a matter neither to be avoided nor regretted.

When the ideas to which I have alluded first occurred to me, it was of course very natural that I should think of M. Valdemar. I knew the steady philosophy of the man too well to apprehend any scruples from *him;* and he had no relatives in America who would be likely to interfere. I spoke to him frankly upon the subject; and, to my surprise, his interest seemed vividly excited. I say to my surprise; for, although he had always yielded his person freely to my experiments, he had never before given me any tokens of sympathy with what I did. His disease was of that character which would admit of exact calculation in respect to the epoch of its termination in death; and it was finally arranged between us that he would send for me about twenty-four hours before the period announced by his physicians as that of his decease.

It is now rather more than seven months since I received, from M. Valdemar himself, the subjoined note:

> MY DEAR P——
> You may as well come now. D—— and F—— are agreed that I cannot hold
> out beyond to-morrow midnight; and I think they have hit the time very nearly.
> VALDEMAR

I received this note within half an hour after it was written, and in fifteen minutes more I was in the dying man's chamber. I had not seen him for ten days, and was appalled by the fearful alteration which the brief interval had wrought in him. His face wore a leaden hue; the eyes were utterly lustreless; and the emaciation was so extreme, that the skin had been broken through by the cheek-bones. His expectoration was excessive. The pulse was barely perceptible. He retained, nevertheless, in a very remarkable manner, both his mental power and a certain degree of physical strength. He spoke with distinctness—took some palliative medicines without aid—and, when I entered the room, was occupied in penciling memoranda in a pocket-book. He was propped up in the bed by pillows. Doctors D—— and F—— were in attendance.

After pressing Valdemar's hand, I took these gentlemen aside, and obtained from them a minute account of the patient's condition. The left lung had been for eighteen months in a semi-osseous or cartilaginous state, and was, of course, entirely useless for all purposes of vitality. The right, in its upper portion, was also partially, if not thoroughly, ossified, while the lower region was merely a mass of purulent tubercles, running one into another. Several extensive per- forations existed; and, at one point, permanent adhesion to the ribs had taken place. These appearances in the right lobe were of comparatively recent date. The ossification had proceeded with very unusual rapidity; no sign of it had been discovered a month before, and the adhesion had only been observed during the three previous days. Independently of the phthisis, the patient was suspected of aneurism of the aorta; but on this point the osseous symptoms rendered an exact diagnosis impossible. It was the opinion of both physicians that M. Valdemar would die about midnight on the morrow (Sunday). It was then seven o'clock on Saturday evening.

On quitting the invalid's bedside to hold conversation with myself, Doc- tors D—— and F—— had bidden him a final farewell. It had not been their intention to return; but, at my request, they agreed to look in upon the patient about ten the next night.

When they had gone, I spoke freely with M. Valdemar on the subject of his approaching dissolution, as well as, more particularly, of the experiment proposed. He still professed himself quite willing and even anxious to have it made, and urged me to commence it at once. A male and a female nurse were in attendance; but I did not feel myself altogether at liberty to engage in a task of this character with no more reliable witnesses than these people, in case of sudden accident, might prove. I therefore postponed operations until about eight the next night, when the arrival of a medical student, with whom I had some acquaintance (Mr. Theodore L——l), relieved me from further embarrassment. It had been my design, originally, to wait for the physicians; but I was induced to proceed, first, by the urgent entreaties of M. Valdemar, and secondly, by my conviction that I had not a moment to lose, as he was evidently sinking fast.

Mr. L——l was so kind as to accede to my desire that he would take notes of all that occurred; and it is from his memoranda that what I now have to relate is, for the most part, either condensed or copied *verbatim.*

It wanted about five minutes of eight when, taking the patient's hand, I begged him to state, as distinctly as he could, to Mr. L——l, whether he (M. Valdemar) was entirely willing that I should make the experiment of mesmerizing him in his then condition.

He replied feebly, yet quite audibly, "Yes, I wish to be mesmerized"—adding immediately afterwards, "I fear you have deferred it too long."

While he spoke thus, I commenced the passes which I had already found most effectual in subduing him. He was evidently influenced with the first lateral stroke of my hand across his forehead; but although I exerted all my powers, no further perceptible effect was induced until some minutes after ten o'clock, when Doctors D—— and F—— called, according to appointment. I explained to them, in a few words, what I designed, and as they opposed no objection, saying that the patient was already in the death agony, I proceeded without hesitation—exchanging, however, the lateral passes for downward ones, and directing my gaze entirely into the right eye of the sufferer.

By this time his pulse was imperceptible and his breathing was stertorous, and at intervals of half a minute.

This condition was nearly unaltered for a quarter of an hour. At the expiration of this period, however, a natural although a very deep sigh escaped from the bosom of the dying man, and the stertorous breathing ceased—that is to say, its stertorousness was no longer apparent; the intervals were undiminished. The patient's extremities were of an icy coldness.

At five minutes before eleven I perceived unequivocal signs of the mesmeric influence. The glassy roll of the eye was changed for that expression of uneasy *inward* examination which is never seen except in cases of sleep-waking, and which it is quite impossible to mistake. With a few rapid lateral passes I made the lids quiver, as in incipient sleep, and with a few more I closed them altogether. I was not satisfied, however, with this, but continued the manipulations vigorously, and with the fullest exertion of the will, until I had completely stiffened the limbs of the slumberer, after placing them in a seemingly easy position. The legs were at full length; the arms were nearly so, and reposed on the bed at a moderate distance from the loins. The head was very slightly elevated.

When I had accomplished this, it was fully midnight, and I requested the gentlemen present to examine M. Valdemar's condition. After a few experiments, they admitted him to be in an unusually perfect state of mesmeric trance.

The curiosity of both the physicians was greatly excited. Dr. D—— resolved at once to remain with the patient all night, while Dr. F—— took leave with a promise to return at daybreak. Mr. L——l and the nurses remained.

We left M. Valdemar entirely undisturbed until about three o'clock in the morning, when I approached him and found him in precisely the same condition as when Dr. F—— went away—that is to say, he lay in the same position; the pulse was imperceptible; the breathing was gentle (scarcely noticeable, unless through the application of a mirror to the lips); the eyes were closed naturally; and the limbs were as rigid and as cold as marble. Still, the general appearance was certainly not that of death.

As I approached M. Valdemar I made a kind of half effort to influence his right arm into pursuit of my own, as I passed the latter gently to and fro above his person. In such experiments with this patient, I had never perfectly succeeded before, and assuredly I had little thought of succeeding now; but to my astonishment, his arm very readily, although feebly, followed every direction I assigned it with mine. I determined to hazard a few words of conversation.

"M. Valdemar," I said, "are you asleep?" He made no answer, but I perceived a tremor about the lips, and was thus induced to repeat the question, again and again. At its third repetition, his whole frame was agitated by a very slight shivering; the eyelids unclosed themselves so far as to display a white line of the ball; the lips moved sluggishly, and from between them, in a barely audible whisper, issued the words:

"Yes;—asleep now. Do not wake me!—let me die so!"

I here felt the limbs, and found them as rigid as ever. The right arm, as before, obeyed the direction of my hand. I questioned the sleep-waker again:

"Do you still feel pain in the breast, M. Valdemar?"

The answer now was immediate, but even less audible than before:

"No pain—I am dying."

I did not think it advisable to disturb him further just then, and nothing more was said or done until the arrival of Dr. F—— who came a little before sunrise, and expressed unbounded astonishment at finding the patient still alive. After feeling the pulse and applying a mirror to the lips, he requested me to speak to the sleep-waker again. I did so, saying:

"M. Valdemar, do you still sleep?"

As before, some minutes elapsed ere a reply was made; and during the interval the dying man seemed to be collecting his energies to speak. At my fourth repetition of the question, he said very faintly, almost inaudibly:

"Yes; still asleep—dying."

It was now the opinion, or rather the wish, of the physicians, that M. Valde-mar should be suffered to remain undisturbed in his present apparently tranquil condition, until death should supervene—and this, it was generally agreed, must now take place within a few minutes. I concluded, however, to speak to him once more, and merely repeated my previous question.

While I spoke, there came a marked change over the countenance of the sleep-waker. The eyes rolled themselves slowly open, the pupils disappearing upwardly; the skin generally assumed a cadaverous hue, resembling not so much parchment as white paper; and the circular hectic spots which, hitherto, had been strongly defined in the centre of each cheek, *went out* at once. I use this expression, because the suddenness of their departure put me in mind of nothing so much as the extinguishment of a candle by a puff of the breath. The upper lip, at the same time, writhed itself away from the teeth, which it had previously covered completely; while the lower jaw fell with an audible jerk, leaving the mouth widely extended, and disclosing in full view the swollen and blackened tongue. I presume that no member of the party then present had been unaccustomed to death-bed horrors; but so hideous beyond conception was the appearance of M. Valdemar at this moment, that there was a general shrinking back from the region of the bed.

I now feel that I have reached a point of this narrative at which every reader will be startled into positive disbelief. It is my business, however, simply to proceed.

There was no longer the faintest sign of vitality in M. Valdemar; and con-cluding him to be dead, we were consigning him to the charge of the nurses, when a strong vibratory motion was observable in the tongue. This continued for perhaps a minute. At the expiration of this period, there issued from the distended and motionless jaws a voice—such as it would be madness in me to attempt describing. There are, indeed, two or three epithets which might be considered as applicable to it in part; I might say, for example, that the sound was harsh, and broken and hollow; but the hideous whole is indescribable, for the simple reason that no similar sounds have ever jarred upon the ear of humanity. There were two particulars, nevertheless, which I thought then, and still think, might fairly be stated as characteristic of the intonation—as well adapted to convey some idea of its unearthly peculiarity. In the first place, the voice seemed to reach our ears—at least mine—from a vast distance, or from some deep cavern within the earth. In the second place, it impressed me (I fear, indeed, that it will be impossible to make myself comprehended) as gelatinous or glutinous matters impress the sense of touch.

I have spoken both of "sound" and of "voice." I mean to say that the sound was one of distinct—of even wonderfully, thrillingly distinct—syllabification. M. Valdemar *spoke*—obviously in reply to the question I had propounded to him a few minutes before. I had asked him, it will be remembered, if he still slept. He now said:

"Yes;—no;—I *have been* sleeping—and now—now—*I am dead.*"

No person present even affected to deny, or attempted to repress, the unutterable, shuddering horror which these few words, thus uttered, were so well calculated to convey. Mr. L——l (the student) swooned. The nurses immediately left the chamber, and could not be induced to return. My own impressions I would not pretend to render intelligible to the reader. For nearly an hour, we busied ourselves, silently—without the utterance of a word—in endeavors to revive Mr. L——l. When he came to himself, we addressed ourselves again to an investigation of M. Valdemar's condition.

It remained in all respects as I have last described it, with the exception that the mirror no longer afforded evidence of respiration. An attempt to draw blood from the arm failed. I should mention, too, that this limb was no further subject to my will. I endeavored in vain to make it follow the direction of my hand. The only real indication, indeed, of the mesmeric influence, was now found in the vibratory movement of the tongue, whenever I addressed M. Valdemar a question. He seemed to be making an effort to reply, but had no longer sufficient volition. To queries put to him by any other person than myself he seemed utterly insensible—although I endeavored to place each member of the company in mesmeric *rapport* with him. I believe that I have now related all that is necessary to an understanding of the sleep-waker's state at this epoch. Other nurses were procured; and at ten o'clock I left the house in company with the two physicians and Mr. L——l.

In the afternoon we all called again to see the patient. His condition remained precisely the same. We had now some discussion as to the propriety and feasibility of awakening him; but we had little difficulty in agreeing that no good purpose would be served by so doing. It was evident that, so far, death (or what is usually termed death) had been arrested by the mesmeric process. It seemed clear to us all that to awaken M. Valdemar would be merely to insure his instant, or at least his speedy dissolution.

From this period until the close of last week—*an interval of nearly seven months*—we continued to make daily calls at M. Valdemar's house, accompanied, now and then, by medical and other friends. All this time the sleep-waker remained *exactly* as I have last described him. The nurses' attentions were continual.

It was on Friday last that we finally resolved to make the experiment of awakening, or attempting to awaken him; and it is the (perhaps) unfortunate result of this latter experiment which has given rise to so much discussion in private circles—to so much of what I cannot help thinking unwarranted popular feeling.

For the purpose of relieving M. Valdemar from the mesmeric trance, I made use of the customary passes. These, for a time, were unsuccessful. The first indication of revival was afforded by a partial descent of the iris. It was observed, as especially remarkable, that this lowering of the pupil was accompanied by the profuse out-flowing of a yellowish ichor (from beneath the lids) of a pungent and highly offensive odor.

It was now suggested that I should attempt to influence the patient's arm, as heretofore. I made the attempt and failed. Dr. F—— then intimated a desire to have me put a question. I did so, as follows:

"M. Valdemar, can you explain to us what are your feelings or wishes now?"

There was an instant return of the hectic circles on the cheeks; the tongue quivered, or rather rolled violently in the mouth (although the jaws and lips remained rigid as before;) and at length the same hideous voice which I have already described, broke forth:

"For God's sake!—quick!—quick!—put me to sleep—or, quick!—waken me!—quick!—*I say to you that I am dead!*"

I was thoroughly unnerved, and for an instant remained undecided what to do. At first I made an endeavor to re-compose the patient; but, failing in this through total abeyance of the will, I retraced my steps and as earnestly struggled to awaken him. In this attempt I soon saw that I should be successful—or at least I soon fancied that my success would be complete—and I am sure that all in the room were prepared to see the patient awaken.

For what really occurred, however, it is quite impossible that any human being could have been prepared.

As I rapidly made the mesmeric passes, amid ejaculations of "dead! dead!" absolutely *bursting* from the tongue and not from the lips of the sufferer, his whole frame at once—within the space of a single minute, or less, shrunk—crumbled—absolutely *rotted* away beneath my hands. Upon the bed, before that whole company, there lay a nearly liquid mass of loathsome—of detestable putridity.

DISCUSSION QUESTIONS

1. If Valdemar provides informed consent to be mesmerized, as one scholar (Altschuler 2003) has argued, how does the informed consent process in the tale compare to the process in clinical research today? What do you believe motivates Valdemar to be mesmerized at the moment of death? How does his contribution to "research" compare with contributions made by patients today? What ethical mandates guide the behavior of clinical and biomedical researchers when dealing with human subjects? What barriers keep patients from being fully "informed" before providing consent?

2. The tale features perhaps the most interprofessional group of providers in any of Poe's tales. How well does this group function? Is a hierarchy in place? Does the tale either promote or debunk any of the professional stereotypes with which you are familiar? As the team-based model receives increasing focus in health care, what measures should educators take to ensure all team members know how to communicate effectively and empathetically with other team members and with patients?

3. The medical student, Mr. L——, works closely with the mesmerist-doctor. If we could argue that he learns from a "hidden curriculum"—from what he observes his role models do—what, then, are the lessons he learns? Do you believe role models have a powerful influence on how future providers will behave? What measures might a medical college or academic health science institution take to ensure future providers learn to be not only technically skilled but also empathetic, compassionate, and skilled in communication?

4. What values regarding palliative and end-of-life care are implicit in the tale, and how do they speak to debates related to the ethics of end-of-life care today, including those of hospice care and pain management? Should a patient have the right to choose how he or she will die, particularly in cases of terminal illness in which patients seek the help of physicians for euthanasia?

5. Read Hawthorne's "Rappaccini's Daughter" or "The Birthmark" and compare Hawthorne's treatment of the subject of human experimentation to Poe's here. In particular, consider issues of full disclosure and informed consent as well as issues related to gender and power.

Notes

1. Physician Oliver Wendell Holmes, professor of anatomy and physiology at Dartmouth College and later dean at Harvard University, delivered a lecture titled "Homeopathy and Its Kindred Delusions" in 1842. An important nineteenth-century poet and writer himself, Holmes was a friend to Nathaniel Hawthorne. When Poe died, Holmes was one of a few writers to send a letter to be read at his funeral.

2. Dr. Valentine Mott was a prominent New York surgeon and professor of Columbia College (now Columbia University). In 1836, Poe wrote a review of William Leete Stone's *Ups and Downs in the Life of a Distressed Gentleman. By the Author of Tales and Sketches, Such as They Are,* which features a young Doctor Wheelwright, who "reads the first chapter of *Cheselden's Anatomy,* visits New York, attends the lectures of Hosack and Post, 'presses into his goblet the grapes of wisdom clustering around the tongue of Mitchill, and acquires the principles of surgery from the lips, and the skilfull use of the knife from the untrembling hand, of Mott.'" The other doctors Leete refers to are William Cheselden, British physician and surgeon, who wrote *Osteographia; or, The Anatomy of Bones,* the first comprehensive and accurate examination of the human skeletal system; and David Hosack (1769–1835), Alfred Charles Post (1806–1886), and Samuel L. Mitchill (1764–1831), prominent nineteenth-century American physicians and lecturers.

3. These fragments were published in *The Bookman* (1909) and can be found in *The Collected Works of Edgar Allan Poe,* vol. 1, *Poems* (1969), edited by Thomas Ollive Mabbott.

4. John J. Moran was the resident physician of Washington College Hospital to whose care Poe was entrusted October 3, 1849, by Poe's friend Joseph Evans Snodgrass and uncle by marriage, Henry Herring. Moran's accounts of Poe's condition and his death were inconsistent over the years, as were Snodgrass's. Most egregiously, Snodgrass seems to have used Poe's death as a cautionary tale for the temperance movement. Snodgrass received a note from Joseph W. Walker, who found Poe in a Baltimore tavern, which read,

Dear Sir:—
There is a gentleman rather the worse for wear, at Ryan's Fourth Ward polls, who goes under the cognomen of Edgar A. Poe, and who appears in great distress. He says that he is acquainted with you, and I assure you he is in need of immediate assistance. Yours, in haste,

Jos. W. Walker.

Snodgrass later misquoted this letter, claiming Walker had written that Poe was "in a state of beastly intoxication." For more information, consult Snodgrass (1867), Bandy (1987), and Walsh (2000).

5. Scott Peeples's *Afterlife of Edgar Allan Poe* (2013) offers a comprehensive study of these readings.

6. The line between fiction and nonfiction was often blurred during this era, as many magazines and newspapers printed tales and news features side by side, without clearly distinguishing between the two. This blurring of lines allowed Poe to trick readers into believing tales like "The Balloon Hoax" (title assigned later, not included with first publication) and "The Facts in the Case of M. Valdemar" were factual reports.

7. See Poe's review of "Phrenology, and the Moral Influence of Phrenology: Arranged for General Study, and the Purposes of Education, from the first Published Works of Gall and Spurzheim, to the Latest Discoveries of the Present Period. By Mrs. L. Miles," in *Complete Works,* edited by Harrison (Poe 1902).

8. See Poe's April 5, 1845, review of "Human Magnetism; Its Claim to Dispassionate Inquiry. Being an Attempt to Show the Utility of Its Application for the Relief of Human Suffering. By W. Newnham, Esq., M.R.S.L., Author of the 'Reciprocal Influence of Body and Mind,' Etc.," in *Complete Works,* edited by Harrison (Poe 1902).

9. In colonial America, mental illnesses were often lumped into two categories: mania or melancholy. In the early 1800s, "monomania" emerged as a diagnostic term encompassing various types of obsessive behavior that are today diagnosed as distinct mental disorders.

10. See, as examples, July 20, 1837, June 16, 1838, August 11, 1838, and December 19, 1850, in *Passages from the American Note-Books* (Hawthorne 1886).

11. Poe's tale "Ligeia" also mentions physicians who never appear in the tale. Like Madeleine Usher's doctor in "The Fall of the House of Usher," Ligeia's physicians are unable to diagnose and treat her, and the only medicine she has apparently been prescribed is wine. I selected "Berenice" and "The Fall of the House of Usher" instead of "Ligeia" because I thought these tales offered more opportunities to discuss the doctor's role in a meaningful way.

THE HAUNTED QUACK

1. Lydia Pinkham was a Massachusetts resident who developed tonics to relieve women of menopausal and menstrual suffering. She has been credited with offering women healing options during a time when mainstream medicine offered little to no relief other than surgery, which had a high mortality rate (Harvard University Library 2015). Because she was so successful at mass marketing her products, however, she has also been compared to the average snake-oil salesman.

2. In 1830, Hawthorne wrote Samuel Goodrich, editor of the *Token,* that he was sending two new pieces. One of these was "Sights from a Steeple." Although published under the name Joseph Nicholson, "The Haunted Quack" is, according to Arlin Turner (1980), the "most likely" candidate for the second piece. Turner notes that "The Haunted Quack" was written with the "care and finish of language usual with [Hawthorne]" and that it included a trip down the Erie Canal, which Hawthorne "may have taken by 1830" (1980, 404). The tale is included in Lathrop's *Complete Works.*

3. Dr. William Buchan (1729–1805) was author of *Domestic Medicine,* a book first published in London in 1769 and reprinted several times in America.

4. Herman Boerhaave (1668–1738) was a Dutch botanist and physician known for promoting a lesion-based model of medicine and bedside teaching. Boerhaave syndrome is named for him.

5. This is the eponymous character from Christopher Marlowe's play *The Tragic History of the Life and Death of Doctor Faustus* (published 1604). Based on the older legend of Faust, the play depicts the life of a necromancer-doctor who makes a pact with the devil.

THE MINISTER'S BLACK VEIL

1. Similar to Ostrowski (1998), who proposes this syphilis theory for "The Minister's Black Veil," Bensick (1985) argues in *La Nouvelle Beatrice* that Beatrice and others in "Rappaccini's Daughter" are infected with syphilis.

LADY ELEANORE'S MANTLE

1. Poe's accusation is groundless because Hawthorne's "Howe's Masquerade" (1838) was published before "William Wilson" (1839). For additional discussion of how Poe and Hawthorne influenced and borrowed from one another, see Herndon (2006), Kopley (2003), and Regan (1970).

2. For a discussion of the political context of Hawthorne's tale, see Lee (2004) and Gross (1955).

THE BIRTHMARK

1. For more discussion of women's health in the nineteenth century, see Smith-Rosenberg (1972), Smith-Rosenberg and Rosenberg (1973), Browner (2005), Welter (1966), Wendland (2006), and Schwartz (2006).

2. Hawthorne also refers to Paracelsus in *The Scarlet Letter,* and Bensick describes the influence of Paracelsus on "Rappaccini's Daughter" in *La Nouvelle Beatrice* (1985).

3. The *Philosophical Transactions* of the Royal Society of London was the first scientific journal; its first issue was published March 6, 1665.

EGOTISM; OR, THE BOSOM SERPENT

1. In "Hawthorne's Literary Borrowings," Arlin Turner (1936) argues that Southey's "Roderick, the Last of the Goths" is a source for this tale as well. Southey's poem features a character named Roderick, "driven by solitude and a persecuting conscience" (553), and Turner suggests Hawthorne alludes to Southey's Roderick in a passage of "Egotism."

RAPPACCINI'S DAUGHTER

1. Browner (2005) notes in *Profound Science* that prior to the work of Vesalius, the most prominent Paduan anatomist, the interior of the body was considered "the soul's domain" (58); for this reason, the church banned human dissection and students instead learned from the Greek physician Galen's teachings. At the same time, Bensick (1985) argues that "Rappaccini's Daughter" can be read as a rivalry between traditional Galenists and practitioners like Paracelsus, who promoted observation and experimentation. Stripling (2013) provides a synopsis of this Galen-Paracelsus theory in *Bioethics and Medical Issues.*

2. For more discussion of women's health in the nineteenth century, see Smith-Rosenberg (1972), Smith-Rosenberg and Rosenberg (1973), Browner (2005), Welter (1966), Wendland (2006), and Schwartz (2006).

BERENICE

1. In two 1835 letters to *Southern Literary Messenger* publisher Thomas Willis White, Poe refers to a Dr. Buckler and his wife, Eliza Sloan Buckler (Poe 1835). John D. Buckler was a Baltimore physician who had been treating Poe for an illness. Eliza Sloan Buckler was a poet who published a poem titled "Answer" in the *Southern Literary Messenger* in 1835 in response to Richard Henry Wilde's "My Life Is Like a Summer Rose." For a comprehensive collection of Poe's letters, see *The Collected Letters of Edgar Allan Poe* (2008).

2. For more discussion of women's health in the nineteenth century, see Smith-Rosenberg (1972), Smith-Rosenberg and Rosenberg (1973), Browner (2005), Welter (1966), Wendland (2006), and Schwartz (2006). For a discussion of hysteria in "Berenice," see Renzi (2012).

3. The epigraph to "Berenice" translates as "My companions told me I might find some little alleviation from my misery, in visiting the grave of my beloved."

4. The sentence from Tertullian's "De Carne Christi" translates as "The Son of God is dead; it is credible because it is absurd. And, buried, He rose again: it is certain, because it is impossible."

5. "*Que tous ses pas étaient des sentiments,*" from Mlle. Marie Salle (1714–1756), translates as "Her every step was a sentiment." The second quote, "*que toutes ses dents étaient des idées. Des idées!*" translates as "that all her teeth were ideas. Ideas!"

THE FALL OF THE HOUSE OF USHER

1. For a comprehensive analysis of incest in nineteenth-century America, including a more nuanced discussion of physiologists' and phrenologists' perceptions of incest and familial eroticism, see Connolly (2014).

2. Scholars have debated whether there is clear evidence of incest in "The Fall of the House of Usher." I concur with Scott Peeples (1998), who argues that the evidence of incest is "hard to deny" (85).

3. The epigraph to the tale translates as "His heart is a lute strung tight; As soon as one touches it, it resounds." The lines are from "Le Refus" (1831), by Pierre-Jean de Béranger. The original read, "My heart," which Poe changed to "His heart."

THE BLACK CAT

1. The June 1875 letter written by Mary Louise Barney Shew Houghton about her diagnosis of a brain lesion can be found in Thomas and Jackson's 1987 *The Poe Log.*

MESMERIC REVELATION

1. Townshend's book, first published in London in 1840, was reprinted in Boston (1841), New York (1841 and 1842), and again in London (1843 and 1844). Lind (1947) argues that Poe consulted the London (1844) edition for "Mesmeric Revelation" and "The Facts in the Case of M. Valdemar." In a review of a book by W. Newnham titled *Human Magnetism; Its Claim to Dispassionate Inquiry,* Poe disagrees with Newnham's "disparagement of the work of Chauncey Hare Townshend, which we regard as one of the most truly profound and philosophical works of the day—a work to be valued properly only in a day to come" (Poe 1984, 123)

2. In a July 2, 1844, letter to James Russell Lowell, Poe espoused the metaphysical views Vankirk describes in "Mesmeric Revelation." As Ostrom and colleagues note in *The Letters* (2008), the phrasing in Poe's letter is strikingly similar to that in the tale. Eight days later, in a July 10, 1844, letter to Dr. Thomas H. Chivers, Poe writes the following: "My own faith is indeed my own. You will find it, somewhat detailed, in a forthcoming number of the 'Columbian Magazine', published here. I have written for it an article headed 'Mesmeric Revelation,' which see. It may be out in the August or September number." See Poe (2008). In August 1844, Poe sent Lowell a copy of the *Columbian Magazine* containing "Mesmeric Revelation." In the letter accompanying the magazine, he inquires of Lowell, "Do you ever see Mr. Hawthorne? He is a man of rare genius. A day or two since I met with a sketch by him called 'Drowne's Wooden Image'—delicious."

SOME WORDS WITH A MUMMY

1. In *The Collected Works* (Poe 1969), editor Mabbott suggests that in 1823–24, the young Poe was likely taken by his foster parents, the Allans, to see the Egyptian mummy exhibit at the Senate Chamber of the Capitol in Norfolk, Virginia.

2. Most famous of Poe's tales involving code-breaking is "The Goldbug," set on Sullivan's Island, South Carolina, where Poe (under the alias Edgar A. Perry) was stationed at Fort Moultrie from November 1827 to December 1828. Charleston legend has it that Poe befriended Edmund Ravenel, a local physician and teacher at the Medical College of South Carolina, who was also a conchologist; however, while biographer Quinn (1998) proposes it is likely that Poe would have met Ravenel, who also served as physician at Fort Moultrie during the time Poe was stationed there, Mabbott (1969) argues that there is no real evidence of their acquaintance. Needless to say, then, the ghost stories in which Poe (Edgar Perry) is said to have fallen in love with Dr. Ravenel's daughter, Anne, who allegedly haunts a local graveyard, are also unsubstantiated.

THE SYSTEM OF DOCTOR TARR AND PROFESSOR FETHER

1. In 1840, Dr. Pliny Earl took an interest in helping Poe establish his own *Penn Magazine*, and Poe wrote to thank him for his support (*Letters* 2008) and to report that he would publish one of Dr. Earl's poems, "By an Octogenarian," in the first run of the magazine. Poe's dream of publishing the magazine in January 1841 was never realized. He continued to pursue the establishment of his own magazine, later under the name the *Stylus*, but he was never successful in this regard. Illness and poverty both played a role in this failure.

THE FACTS IN THE CASE OF M. VALDEMAR

1. Hawthorne's "The Birthmark" arguably provides an earlier example of informed consent, but Altschuler's point is that the consent in "Valdemar" describes an official consent process: the doctor explains his research plans to the patient and receives his verbal consent in front of a witness. "The Birthmark," in fact, might be read as an example of coercion or undue influence.

References

WORKS CITED

Altschuler, Eric Lewin. 2003. "Informed Consent in an Edgar Allan Poe Tale." *Lancet* 362 (9394): 1504.

Altschuler, Eric Lewin, and Seth Augenstein. 2012. "The Earliest Description of the Frontal Lobe Syndrome in an Edgar Allan Poe Tale." *Brain Injury* 26 (11): 1403–4.

Amper, Susan. 1992. "Untold Story: The Lying Narrator in 'The Black Cat.'" *Studies in American Fiction* 29 (4): 475–85.

Aull, Felice. 1993. "Literature, Arts, and Medicine Database." NYU School of Medicine, Langone Medical Center Web site. Last modified 2015. http://medhum.med.nyu.edu/index.html.

Bandy, William T. 1987. "Dr. Moran and the Poe-Reynolds Myth." *Myths and Reality,* edited by Benjamin Franklin Fisher IV, 26–36. Baltimore: Edgar Allan Poe Society. http://www.eapoe.org/papers/psbbooks/pb19871d.htm.

Battle, Robert W. 2011. "Edgar Allan Poe: A Case Description of the Marfan Syndrome in an Obscure Short Story." *American Journal of Cardiology* 108 (1): 148–49.

Baym, Nina. 1982. "Thwarted Nature: Nathaniel Hawthorne as Feminist." In *American Novelists Revisited: Essays in Feminist Criticism,* edited by Fritz Fleischmann, 58–64. Boston: G. K. Hall.

———, ed. 2003. *The Norton Anthology of American Literature Shorter Version,* 6th ed. New York: Norton.

———. 2005. "Revisiting Hawthorne's Feminism." In *Nathaniel Hawthorne and the Real: Bicentennial Essays,* edited by Millicent Bell, 107–24. Columbus: Ohio State University Press.

Benjamin, Ludy T. Jr., and David B. Baker. 2004. *From Séance to Science: A History of the Profession of Psychology in America.* Belmont, California: Thomson/Wadsworth Learning.

Bensick, Carol Marie. 1985. *La Nouvelle Beatrice: Renaissance and Romance in "Rappaccini's Daughter."* New Brunswick, New Jersey: Rutgers University Press.

Benton, Richard P. 1968. "Poe's 'The System of Dr. Tarr and Prof. Fether': Dickens or Willis?" *Poe Newsletter* 1 (1): 7–9.

Browner, Stephanie. 1993. "Authorizing the Body: Scientific Medicine and *The Scarlet Letter.*" *Literature and Medicine* 12 (2): 139–60.

———. 2005. *Profound Science and Elegant Literature: Imagining Doctors in Nineteenth-Century America.* Philadelphia: University of Pennsylvania Press.

Bush, Sargent, Jr. 1971. "Bosom Serpents before Hawthorne: The Origins of a Symbol." *American Literature* 43 (2): 181–99.

Campbell, Killis. 1933. *The Mind of Poe and Other Studies*. Cambridge, Massachusetts: Harvard University Press.

Cassedy, James H. 1991. *Medicine in America: A Short History*. Baltimore: Johns Hopkins University Press.

Cerulli, Anthony, and Sarah L. Berry. 2014. "Nathaniel Hawthorne's Warring Doctors and Meddling Ministers." *Mosaic* 47 (1): 111–27.

Charon, Rita. 2006. *Narrative Medicine: Honoring the Stories of Illness*. New York: Oxford University Press.

Connolly, Brian. 2014. *Domestic Intimacies: Incest and the Liberal Subject in Nineteenth-Century America*. Philadelphia: University of Pennsylvania Press.

DasGupta, Sayantani. 2008. "Narrative Humility." *Lancet* 371 (9617): 980–81.

DeBakey, Lois. 1968. "The Fictional Physician-Scientist of Nineteenth-Century America." *Anesthesia and Analgesia* 47 (2): 108–18.

Dibble, Christopher. 2010. "The Dead Ringer: Medicine, Poe, and the Fear of Premature Burial." *Historia Medicinae* 2 (1): 2–9.

Dolezal, J. 2005. "The Medical Palimpsest of *The Scarlet Letter*: An Interdisciplinary Reading." *Medical Humanities* 31 (1): 17–22.

Elbert, Monika. 2004a. "Poe and Hawthorne as Women's Amanuenses." *Poe Studies/Dark Romanticism* 37 (1–2): 21–37.

———. 2004b. "The Surveillance of Woman's Body in Hawthorne's Short Stories." *Women's Studies* 33 (1): 23–46.

Engel, John D. et al., eds. 2008. *Narrative in Health Care: Healing Patients, Practitioners, Profession, and Community*. Oxford: Radcliffe.

Falk, Doris. 1969. "Poe and the Power of Animal Magnetism." *PMLA* 84 (3): 536–46.

Fetterly, Judith. 1978. *The Resisting Reader: A Feminist Approach to American Fiction*. Bloomington: Indiana University Press.

Fletcher, Joann. 2011. "Mummies around the World." British Broadcast System Web site. http://www.bbc.co.uk/history/ancient/egyptians/mummies_01.shtml.

Forclaz, Roger. 1968. "A Source for 'Berenice' and a Note on Poe's Reading." *Poe Newsletter* 1 (2): 25–27.

Ford, Dana. 2014. "Mississippi Man Who Awoke in Body Bag Dies Two Weeks Later." CNN Web site. http://www.cnn.com/2014/03/13/us/mississippi-walter-williams-dead/.

Foucault, Michel. 1994. *The Birth of the Clinic: An Archaeology of Medical Perception*. New York: Vintage.

Furst, Lillian R., ed. 2000. *Medical Progress and Social Reality: A Reader in Nineteenth-Century Medicine and Literature*. New York: State University of New York Press.

Geiling, Natasha. "The (Still) Mysterious Death of Edgar Allan Poe." *Smithsonian Magazine*, October 7, 2014. Accessed February 24, 2015. http://www.smithsonianmag.com/history/still-mysterious-death-edgar-allan-poe-180952936/.

Goldman, Eric. 2004. "Explaining Mental Illness: Theology and Pathology in Nathaniel Hawthorne's Short Fiction." *Nineteenth-Century Literature* 59 (1): 27–52.

Goodwin, C. James. 1999. *A History of Modern Psychology*, 4th ed. New York: Wiley.

Gross, Seymour L. October 1955. "Hawthorne's 'Lady Eleanore's Mantle' as History." *The Journal of English and Germanic Philosophy* 54 (4): 549–54.

Hardy, Sarah Boykin. 1998. "The Art of Diagnosis." *Narrative* 6 (2): 157–73.

Harvard University Library Open Collections Program. 2015. "Lydia Estes Pinkham (1819–1883)." *Women Working: 1800–1930*. Accessed February 28, 2015. http://ocp.hul.harvard.edu/ww/pinkham.html.

Hawthorne, Manning. 1939. "The Friendship between Hawthorne and Longfellow." *English Journal* 28 (3): 221–23.

Hawthorne, Nathaniel. 1883. *The Complete Works of Nathaniel Hawthorne,* edited by George Parsons Lathrop. Boston: Houghton, Mifflin.

———. 1886. *Passages from the American Note-Books.* Boston, Massachusetts: Houghton, Mifflin.

Herndl, Diane Price. 1993. *Invalid Women: Figuring Feminine Illness in American Fiction and Culture, 1840–1940.* Chapel Hill: University of North Carolina Press.

Herndon, Jerry A. 2006. "The Masque of the Red Death: A Note on Hawthorne's Influence." *Masques, Mysteries, and Mastodons: A Poe Miscellany,* edited by Benjamin F. Fisher, 38–44. Baltimore: Edgar Allan Poe Society.

Hoffman, Daniel. 1998. *Poe Poe Poe Poe Poe Poe Poe.* Louisiana State University Press: Baton Rouge.

Hungerford, Edward. 1930. "Poe and Phrenology." *American Literature* 2 (3): 209–31.

Johnson, Charles P. 1844. *A Treatise of Animal Magnetism.* New York: Burgess & Stringer, 1844.

Kennedy, Gerald J. 1977. "Poe and Magazine Writing on Premature Burial." *Studies in the American Renaissance 1977,* edited by Joel Myerson, 165–78. Boston: Twayne.

———. 2001. "Introduction: Poe in Our Time." *A Historical Guide to Edgar Allan Poe,* edited by Gerald J. Kennedy. New York: Oxford University Press.

King, Lucille. 1930. "Notes on Poe's Sources." *University of Texas Studies in English* 10, 128–34.

Kopley, Richard. 2003. *The Threads of the Scarlet Letter: A Study of Hawthorne's Transformative Act.* Newark: University of Delaware Press.

Lathrop, George Parsons. 2009. "Introductory Note to *The Dolliver Romance.*" In *The Dolliver Romance, Fanshawe, and Septimius Falcon,* edited by George Parsons Lathrop, 4–7. Cambridge: Cambridge Scholars Publishing.

Lee, Sohul. 2004. "Hawthorne's Politics of Storytelling: 'Two Tales of the Province House' and the Specter of Anglomania in the *Democratic Review.*" *American Periodicals* 14 (1): 35–62.

Levine, Stuart, and Susan Levine. 1990. *The Short Fiction of Edgar Allan Poe: An Annotated Edition.* Urbana: University of Illinois Press.

Lind, Sidney E. 1947. "Poe and Mesmerism." *PMLA* 62 (4): 1077–94.

Long, John C. 1989. "The Scene at the Sickbed: Poe, Hawthorne, and Whitman—The Clinic as Discourse in Tales and Poems of Morbid Physic." *University of Hartford Studies in Literature* 21: 21–37.

Matheson, T. J. 1986. "Poe's 'The Black Cat' as a Critique of Temperance Literature." *Mosaic* 19 (3): 69–81.

Mellow, James R. 1980. *Nathaniel Hawthorne in His Times.* Baltimore: Johns Hopkins University Press.

Meyers, Jeffrey. 1992. *Edgar Allan Poe: His Life and Legacy.* New York: Cooper Square Press.

Miller, Edwin Haviland. 1992. *Salem Is My Dwelling Place: A Life of Nathaniel Hawthorne.* Iowa City: University of Iowa Press.

Millington, Richard H. "The Meanings of Hawthorne's Women." Last modified December 10, 2014. http://www.hawthorneinsalem.org/page/10482/.

Montgomery, Travis. 2012. "The Near East." In *Edgar Allan Poe in Context,* edited by Kevin J. Hayes, 53–62. New York: Cambridge University Press.

Ostrowski, Carl. 1998. "The Minister's 'Grievous Affliction': Diagnosing Hawthorne's Parson Hooper." *Literature and Medicine* 17 (2): 197–211.

Peeples, Scott. 1998. *Edgar Allan Poe Revisited.* New York: Twayne.

———. 2003. *The Afterlife of Edgar Allan Poe.* New York: Camden House.

Poe, Edgar Allan. 1835. "Edgar Allan Poe Letter to Thomas W. White—April 30, 1835." Edgar Allan Poe Society of Baltimore Web site. http://www.eapoe.org/works/letters/ p3504300.htm.

———. 1902. *The Complete Works of Edgar Allan Poe,* edited by James A. Harrison. New York: T. Y. Crowell & Company.

———. 1978. *The Collected Works of Edgar Allan Poe, 3 vols.,* edited by Thomas Ollive Mabbott. Cambridge: Belknap Press of Harvard University Press.

———. 1984. *Edgar Allan Poe: Essays and Reviews,* edited by G. R. Thompson. New York: Library of America.

———. 2008. *The Collected Letters of Edgar Allan Poe,* 3d ed., edited by John Ward Ostrom, Burton R. Pollin, and Jeffrey A. Savoye. New York: Gordian.

"Poe's Death Is Rewritten as a Case of Rabies, Not Telltale Alcohol." *New York Times,* September 15, 1996. Accessed December 18, 2014. http://www.nytimes.com/1996/09/15/us/ poe-s-death-is-rewritten-as-case-of-rabies- not-telltale-alcohol.html.

Quinn, Arthur Hobson. 1998. *Edgar Allan Poe: A Critical Biography.* Baltimore: Johns Hopkins University Press.

Regan, Robert. 1970. "Hawthorne's 'Plagiary'; Poe's Duplicity," *Nineteenth-Century Fiction* 25 (3): 281–98.

Reilly, John. 1993. "A Source for Immuration in 'The Black Cat.'" *Nineteenth-Century Literature* 48 (1): 93–95.

Renzi, Kristen. 2012. "Hysteric Vocalizations of the Female Body in Edgar Allan Poe's 'Berenice.'" *ESQ: A Journal of the American Renaissance* 58 (4): 601–40.

Reynolds, David S. 1988. *Beneath the American Renaissance: The Subversive Imagination in the Age of Emerson and Melville.* Cambridge: Harvard University Press.

Reynolds, Larry J. 2001. *A Historical Guide to Nathaniel Hawthorne.* New York: Oxford University Press.

Sahli, Nancy. 1982. *Elizabeth Blackwell, M.D. (1821–1910): A Biography,* 2d ed. New York: Arno Press.

Savitt, Todd L. 1997. "Black Health on the Plantation: Masters, Slaves, and Physicians." *Sickness and Health in America: Readings in the History of Medicine and Public Health,* 3d ed., edited by Judith Walzer Leavitt and Ronald Leavitt, 352–68. Madison: University of Wisconsin Press.

Schwartz, Marie Jenkins. 2006. *Birthing a Slave: Motherhood and Medicine in the Antebellum South.* Cambridge, Massachusetts: Harvard University Press.

Sloane, David E. E. 1969. "Chapter 01: Poe's Sources of Medical Information," *Early Nineteenth Century Medicine in Poe's Short Stories,* Master of Arts Thesis, Duke University: 2–15. Edgar Allan Poe Society of Baltimore Web site. http://www.eapoe.org/PAPERS/ misc1921/smt1966b.htm.

Smith-Rosenberg, Carroll. 1972. "The Hysterical Woman: Sex Roles and Sex Conflict in Nineteenth-Century America." *Social Research* 39 (4): 652–78.

Smith-Rosenberg, Carroll, and Charles Rosenberg. 1973. "The Female Animal: Medical and Biological Views of Woman and Her Role in Nineteenth-Century America." *Journal of American History* 60 (2): 332–56.

Snodgrass, Joseph Evans. 1867. "The Facts of Poe's Death and Burial." *Beadle's Monthly,* May, 283–87.

Spurzheim, Johann Gaspar. 1825. *Phrenology; or, The Doctrine of the Mind, and of the Relations between Its Manifestations and the Body.* London: William Clowes.

Stoehr, Taylor. 1974. "Physiognomy and Phrenology in Hawthorne." *Huntington Library Quarterly* 37 (4): 355–400.

———. 1978. *Hawthorne's Mad Scientists: Pseudoscience and Social Science in Nineteenth Century Life and Letters.* Hamden, Connecticut: Archon.

Stripling, Mahala Yates. 2013. *Bioethics and Medical Issues in Literature.* San Francisco: University of California Medical Humanities Press.

Studd, John. 2007. "A Comparison of Nineteenth Century and Current Attitudes to Female Sexuality." *Gynecological Endocrynology* 23 (12): 673–81.

Theriot, Nancy. 1990. "Diagnosing Unnatural Motherhood: Nineteenth-Century Physicians and Puerperal Insanity." *American Studies* 26: 69–88.

Thomas, Dwight, and David Jackson. 1987. *The Poe Log: A Documentary Life of Edgar Allan Poe, 1809–1849.* Boston: G. K. Hall.

Thrailkill, Jane. 2006. "*The Scarlet Letter* Romantic Medicine." *Studies in American Fiction* 34 (1): 3–31.

Tresch, John. 2002. "Extra! Extra! Poe Invents Science Fiction!" In *The Cambridge Companion to Edgar Allan Poe,* edited by Kevin J. Hayes, 113–32. Cambridge: Cambridge University Press.

Turner, Arlin. 1936. "Hawthorne's Literary Borrowings." *PMLA* 51 (2): 543–62.

———. 1980. *Nathaniel Hawthorne: A Biography.* New York: Oxford University Press.

Uroff, M. D. 1972. "The Doctors in 'Rappaccini's Daughter.'" *Nineteenth-Century Fiction* 27 (1): 61–70.

Verbrugge, Martha H. 1976. "Women and Medicine in Nineteenth-Century America." *Signs* 1 (4): 957–72.

Verghese, Abraham. 2011. "Treat the Patient, Not the CT Scan." *New York Times,* February 26, 2011. Accessed April 5, 2014. http://nytimes.com/2011/02/27/opinion/27verghese.html?pagewanted=all&_ro=0.

Walsh, John Evangelist. 2000. *Midnight Dreary: The Mysterious Death of Edgar Allan Poe.* New York: Palgrave Macmillan.

Weekes, Karen. 2002. "Poe's Feminine Ideal." In *The Cambridge Companion to Edgar Allan Poe,* edited by Kevin J. Hayes, 148–62. Cambridge: Cambridge University Press.

Welter, Barbara. 1966. "The Cult of True Womanhood: 1820–1860." *American Quarterly* 18 (2/1): 151–74.

Wendland, Claire. 2006. "Gendering American Medicine in the Nineteenth Century." *Pharos* 69 (3): 30–37.

Wineapple, Brenda. 2001. "Nathaniel Hawthorne, 1804–1864: A Brief Biography." In *A Historical Guide to Nathaniel Hawthorne,* edited by Larry Reynolds, 13–45. New York: Oxford.

———. 2003. *Hawthorne: A Life.* New York: Random House.

Zimmerman, Brett. 1992. "'Moral Insanity' or Paranoid Schizophrenia: Poe's 'The Tell-Tale Heart.'" *Mosaic* 25 (2): 40–48.

———. 2001. "Frantic Forensic Oratory: Poe's 'The Tell-Tale Heart.'" *Style* 35 (1): 34–49.

———. 2007. "Sensibility, Phrenology, and 'The Fall of the House of Usher.'" *Edgar Allan Poe Review* 8 (1): 53.

———. 2009. "Poe as Amateur Psychologist: Flooding, Phobias, and Psychosomatics in 'The Premature Burial.'" *Edgar Allan Poe Review* 10 (1): 7–19.

ADDITIONAL RESOURCES

Baym, Nina. 2013. "The Tales of the Manse Period." In *Nathaniel Hawthorne's Tales,* 2d ed., edited by James McIntosh, 494–98. New York: Norton.

Bazil, Carl W. 1999. "Seizures in the Life and Works of Edgar Allan Poe." *Archives of Neurology* 56 (6): 740–43.

Chaney, Sarah. 2011. "'A Hideous Torture on Himself': Madness and Self-Mutilation in Victorian Literature." *Journal of Medical Humanities* 32 (4): 279–89.

Cleman, John. 2001. "Irresistible Impulses: Edgar Allan Poe and the Insanity Defense." In *Bloom's BioCritiques: Edgar Allan Poe,* edited by Harold Bloom, 66–67. Philadelphia: Chelsea House.

Crews, Frederick. 1989. *The Sins of the Fathers: Hawthorne's Psychological Themes.* Berkeley: University of California Press.

Hastings, Louise. 1938. "An Origin for 'Dr. Heidegger's Experiment.'" *American Literature* 9 (4): 403–10.

Hawthorne, Nathaniel. 2013. *Nathaniel Hawthorne's Tales,* 2d ed., edited by James McIntosh. New York: Norton.

Hayes, Kevin J., ed. 2002. *The Cambridge Companion to Edgar Allan Poe.* Cambridge: Cambridge University Press.

Lemay, J. A. Leo. 1982. "Poe's 'The Businessman': Its Contexts and Satire of Franklin's Autobiography." *Poe Studies* 15 (2): 29–37.

Person, Leland S. 2001. "Poe and Nineteenth Century Gender Constructions." In *A Historical Guide to Edgar Allan Poe,* edited by Gerald J. Kennedy, 129–65. New York: Oxford University Press.

Pfister, Joel. 1991. *The Production of Personal Life: Class, Gender, and the Psychological in Hawthorne's Fiction.* Stanford, California: Stanford University Press.

Roche, David. 2009. "The 'Unhealthy' in 'The Fall of the House of Usher': Poe's Aesthetic of Contamination." *Edgar Allan Poe Review* 10 (1): 20–35.

Sontag, Susan. 1978. *Illness as Metaphor and AIDS and Its Metaphors.* New York: Farrar, Strauss & Giroux.

Stashower, Daniel. 2006. *The Beautiful Cigar Girl: Mary Rogers, Edgar Allan Poe, and the Invention of Murder.* New York: Dutton.

Whalen, Terrence. 1999. *Edgar Allan Poe and the Masses.* Princeton, New Jersey: Princeton University Press.

Young, Philip. 1951. "The Earlier Psychologists and Poe." *American Literature* 22 (4): 442–54.